W9-CPB-939

SUDDEN SPRING

SUDDEN SPRING

STORIES *of* ADAPTATION *in a* CLIMATE-CHANGED SOUTH

Rick Van Noy

The University of Georgia Press
Athens

Athens, Georgia 30602
www.ugapress.org

Designed by Erin Kirk New
Set in 10.5 on 14.5 Minion Pro

Most University of Georgia Press titles are
available from popular e-book vendors.

Printed and bound by Thomson-Shore

The paper in this book meets the guidelines for
permanence and durability of the Committee on
Production Guidelines for Book Longevity of the
Council on Library Resources.

Printed in the United States of America
18 19 20 21 22 C 5 4 3 2 1

Library of Congress Cataloging-in-Publication Data

Names: Van Noy, Rick, 1966– author.
Title: Sudden spring : stories of adaptation in a climate-changed South /
Rick Van Noy.
Description: Athens : The University of Georgia Press, c2018. | Includes
bibliographical references and index.
Identifiers: LCCN 2018019210| ISBN 9780820354361 (hardback : alk. paper) |
ISBN 9780820354378 (ebook)
Subjects: LCSH: Southern States—Climate. | Climatic changes—Southern
States. | Coast changes—Southern States. | Coast changes—Gulf Coast
(U.S.) | Saltwater encroachment—Southern States.
Classification: LCC QC984.S83 V36 2018 | DDC 304.2/50975—dc23
LC record available at https://lccn.loc.gov/2018019210

Contents

CHAPTER 1
Tombstones by the Sea: Our Climate Change Commitment
1

CHAPTER 2
Our Best Defense: Shelling the Naval Base, Virginia
14

CHAPTER 3
The Proximity of Far Away: Climate Change Comes to the Alligator, North Carolina
43

CHAPTER 4
Fish out of Water: High Tide in the Lowcountry, South Carolina
58

CHAPTER 5
Ebb-tide Optimism: Ghosts of the Golden Isles, Georgia
78

CHAPTER 6
The Octopus in the Basement: Surreal Matters in the Sunshine State, Florida
99

CHAPTER 7
Springing Back: Resiliency on the Gulf Coast, Louisiana and Texas
129

CHAPTER 8
Take in the Waters: On the Birthplace of Rivers, West Virginia
166

CHAPTER 9
More Ghosts on the Coasts, and the Last Place to Go
193

Acknowledgments
201

Notes
203

Interviews and Correspondence
215

Index
219

SUDDEN SPRING

1

Tombstones by the Sea

Our Climate Change Commitment

Marsh grass waved in the wind, matted by sea and salt. The blue-green sea itself gleamed in the sun, rollers lapping against a shore scattered with wrack and debris. In many ways, the coastal scene was familiar. Evidence of past cultures could be found—in this case arrowheads. Carol Pruitt-Moore, a tall, short-haired woman in her late fifties, said she often found dozens of them as oyster shells crackled under foot, piles called middens, dating back to the time of arrowheads. We scanned the sand to find some, pausing to talk or swat away biting flies. The wind brought up the gassy odor of sea decay and sunbaked mud.

But there was something very different about this scene compared with other coastal communities I had visited. Scattered on the shore was a rusted axle from a trailer, some concrete blocks from a former foundation, ceramic shards, even a bathroom sink. And there were headstones from a graveyard once well inland when the island was abandoned in the 1920s, now toppled over and drooping into the Chesapeake Bay. One of them shared the surname of my host. They were knocked flat by Hurricane Sandy and disturbed more recently by erosion and a rising sea. Carol found five complete skeletons after the hurricane. The forehead of a skull touched the sand, momentarily between each wave, as if to get clear of the sea. "Don't step there," she told me. "There's a leg bone."

It was a grim portent of what climate change could mean for her community and for others in the Southeast and across the planet.

My host had been coming to the island called Uppards, just north of Tangier Island in the Chesapeake Bay, every day for years, noticing small changes unfolding over time. Like her, farmers and gardeners, birders and naturalists across the county have long taken note of changes on the coast and in weather patterns. Leaves bud out earlier and bird migration patterns have changed. Since the early 1900s, about two-thirds of species studied have shifted toward earlier spring blooming, breeding, or migrating. This is true for amphibians, birds, fish, invertebrates, and mammals as well as trees, plants, corals, and plankton. These changes have been observed on every major continent and ocean, according to Camille Parmesan, a professor at the University of Texas at Austin and at Plymouth University in the U.K. who researches phenology, or the biological impacts of climate change. Season creep it is sometimes called, and it's creepy alright.

The USA National Phenology Network, which gathers leaf-out and bloom data along with information about when species migrate and reproduce from across the United States, confirms that the advancing onset of spring and precipitous shifts between warm and cold temperatures are part of ongoing trends. While the network was officially established in the mid-2000s, observations recorded by its contributing scientists and volunteers date back to the 1950s. Some of the longest-running records, which chronicle first leaf growth of honeysuckles and lilacs across the lower forty-eight states, show a noticeable shift since the 1980s. Like the temperatures recorded as part of climate change research, the leaf-out dates show great variability from year to year but the trend is distinct—earlier warmer temperatures and earlier first buds and blooms. The early buds are susceptible to frost, which can mean crop failures. The phenomenon is called "false spring," and we might also call it a "sudden spring." Warming is happening faster than normal, weather is more variable and abrupt.

According to Camille Parmesan, there is little evidence to suggest that species are adapting to extreme temperature swings. Some species are responding to or accommodating these changes, but that does not necessarily mean an evolutionary adaptation. In 2003, Parmesan and Gary Yohe, an economist at Wesleyan University, analyzed records of the geographical ranges of more than seventeen hundred species of plants and animals. They found that their ranges were moving, on average, about four miles per decade toward the poles. Animals and plants were also moving up mountain slopes. But the pace of change is happening too fast for many animals to adapt. Some may have to be physically moved, called "assisted migration," if they are to be saved. More recently, she and other researchers have found that marine plants and animals have moved even faster than terrestrial ones.

What about humans? Can we adapt? Climate change is not just a future threat but a present reality. And it will surely get worse, unless *we* can change. Certainly, most of us can migrate if we have to, though many have noted how the ill effects of climate change may fall disproportionally on the poor. The consensus among scientists studying climate change is that disruptions in what have been considered normal patterns of seasonal temperature and precipitation, set in motion by the buildup of greenhouse gases, are with us for some time to come. Even if there were a pronounced decrease of such emissions worldwide, what is now in the atmosphere would continue to affect global climate change. We have already seen global temperatures rise in the neighborhood of close to 1.1°C (2°F).

We might say there are two kinds of climate change inertia, social and physical. Though we are already seeing obvious warning signs of what is to come, such as melting glaciers, the climate system has some built-in inertia, and the impacts of past human activities will be felt far into the future. Scientists refer to these unavoidable changes as our climate change commitment. Some of the inertia comes from the elevated levels of carbon dioxide and other greenhouse gases already in the atmosphere. If humans were to cease their emissions overnight, the oceans would quickly absorb some of these gases. But the oceans also release gases back to the atmosphere, and the level of greenhouse gases in the atmosphere would not subside back to preindustrial levels for many centuries. "A large fraction of climate change is largely irreversible on human time scales," warns the Fifth Assessment Report from the Intergovernmental Panel on Climate Change (IPCC). Only if human emissions were "strongly negative over a sustained period" and if tree planting and other activities were to sequester far more carbon than humans release would climate change begin to be reversed. At the moment, emissions are still rising.

And what will change? The melting of snow and ice will expose darker patches of water and land that will absorb more of the sun's heat, accelerating global warming and the retreat of ice sheets and glaciers, causing sea level rise. Currents that transport heat within the oceans will be disrupted. Ocean acidification will continue to rise, with unknown effects on marine life. Thawing permafrost and sea beds will release methane, a greenhouse gas. Droughts will trigger vegetation changes and wildfires, releasing carbon. Species will go extinct. Coastal communities will flood.

Changing self-destructive behaviors can be extremely difficult, as any dieter knows, and unrealistic optimism can be just as counterproductive as pessimism. It can be difficult to change if we think we will never get there and if we think things will eventually work out, that we will find a solution or things will naturally stabilize. Some people think the problems of climate change are something we might postpone until more pressing priorities, such as the economy, are addressed. Such a delay of action may indeed be the biggest obstacle to actions and policies that would slow the rate of global warming and avoid its worst impacts. Even though scientists have repeatedly emphasized the urgency of the situation, their message is not getting through to the general public or to legislators who could make a difference.

Every time the IPCC issues a report (there have now been five assessments), the news picks it up in a way that warns of "dire consequences," as *National Geographic* had it, or some apocalyptic or catastrophic scenario. People experience a kind of dissonance when they hear this: *The world will end, driving is causing this, I drive, so I'm causing the world to end.* It's like being told you're a bully when you know you are not. Bullying has happened, you're against bullying, and you want somebody to do something about the bullies. But who will go first?

One main shift that might occur is telling stories about the people who are going first, those making the change materialize, focusing on opportunities and solutions. A better way to create engagement is less "you're doing it all wrong" than "here's what we might do." People are motivated to make wise choices more by hope and opportunity than by fear, cynicism, and despair. We need to hear more of the communities who are creating a better quality of life for their citizens, growing sustainably, and maintaining a healthy environment. The more people start seeing those who are creating a better society, the sooner they can start taking action.

Perhaps what is missing are vivid, personalized, and local depictions of what life already looks like to communities facing climate change, those with front row seats to the effects. This is not another book on the fact that climate change is happening, nor that we may be reaching an irreversible and apocalyptic "tipping point," nor is it expressly about the politics of climate change. Instead, I focus on communities who are facing up to the inevitable changes they will experience. I look at particular places and the ways its people and communities understand, cope with, and adapt to a changing climate. Just as there are no "atheists in foxholes," I hypothesized that there would be few climate change skeptics in low-lying areas, but I have not found that to be true. Not in Florida, where the very words "climate change" seem to have been struck from official reports. Not on tiny Tangier Island. Still, they will call it whatever you want them to call it if they can get the help they need.

This might have been a different book had the 2016 presidential election turned out differently. But even before the Trump administration said he would remove the United States from the Paris Agreement, or roll back the Climate Action Plan, I wanted to focus on what we need to prepare for. Skeptical readers may be interested in what I have found are some of the sources of this skepticism, including our own inability to process information analytically. They may also be interested in some of the many leaders in Southern communities who are working to disentangle climate change from partisanship, including Republicans Bob Inglis of South Carolina, Mayor Jim Cason of Coral Gables, Florida, and Mayor Jason Buelterman of Tybee Island, Georgia. A Trump presidency is a setback to efforts to curb greenhouse gas emissions, but it will not change atmospheric chemistry.

While some have doubts about the causes of climate change, there are few doubts about the effects in those same places, even when recast as "recurrent coastal flooding" or "nuisance flooding." In some places, the community jumps over the political debate and plans for the future. By focusing on the stories of the places making changes, I do not believe we can ignore the damage. We must prepare for both immediate and lasting relief.

For those wishing to catch up on the science, here is an overview. The main cause of global warming is carbon dioxide in the atmosphere, caused by the burning of fossil fuels. Carbon dioxide traps heat—without it the Earth would be covered in ice. In March 2016, the concentration of carbon dioxide climbed to over four hundred parts-per-million (ppm) and continues to increase at about three ppm per year. This amounts to about 50 percent more than it was before the Industrial Revolution. The last time the concentration was this high was several million years ago. Most of the extra heat energy is stored in the ocean and released later, so at present, we are seeing the effects of past excess. The recent high temperatures, extreme storms, and prolonged droughts are small hints of things to come. No one under thirty reading this has ever experienced a below-average month of temperatures. The heating of the oceans has caused them to expand, fueling part of the rise. Some also comes from the melting of polar ice, which will worsen. Scientists have found that the sea rose only about 1.1 millimeters per year before 1990, whereas in the period between 1993 through 2012, it rose 3.1 millimeters per year. NASA puts the present rate of sea level rise at 3.4 millimeters per year. It is not just rising; it is rising at a faster rate.

John Holdren, President Obama's science advisor and formerly an energy expert at Harvard, said that "we basically have three choices: mitigation, adaptation and suffering. We're going to do some of each. The question is what the mix is going to be. The more mitigation we do, the less adaptation will be required and the less suffering there will be." Holdren has spoken of the need to reduce carbon emissions but to make parallel efforts in adaptation, such things as managing coastal development, creating drought resistant crops, and improving the efficiency of water use. Such a commitment would take leadership, coordination, and funding.

The mitigation aspect to Holdren's trinity is not going very well. Though the 2015 Paris Agreement will help globally, at present, humans are failing to put the brakes on the rise of carbon emissions. With the United States pulling out of the agreement, the situation will worsen, but the same basic fact remains—we will still have to adapt. We will have to make decisions about how we will live with coming changes already set in motion. Adaptation involves making changes to activities, rules, and institutions to minimize risk and damage.

The whole idea of natural selection rests on the gradual adaptation of organisms to their environment. If the environment changes, some individuals will have characteristics that make them better able to cope than others, and these individuals will be more likely to pass on their genes to future generations. Some cities are poised to lead into the next generation, as some countries, such as China, will surely fill the hole left by our pulling out of the Paris Agreement. One city official said to me that at some point the federal government will come around to planning and funding for communities to get ready for the changes that will come. Those places that have the design plans ready will be poised to lead.

Adaptation is the evolutionary process whereby an organism becomes better able to live in its habitat. Ironically, it may be features of our own evolution that are causing the delay. We have evolved to respond to immediate threats and risks, not far off ones. In the places I visit, I highlight some of the species that are under stress and the adaptations they have developed to survive, which I hope sheds some oblique light on our own situation. I never cease to be struck by the beauty and vitality of these creatures, despite their being imperiled. Perhaps their very lack of knowledge about the situation is so inspiring. They just live, sing, roar. Despite the decline of many, so much still pulses with life. The paradox that emerged: the trends point to some calamities for much that blooms, migrates, and moves, yet in them is also the source of comfort, sanity, and wonder to carry on.

Just as there will be ripple effects in natural systems, feedback loops that amplify other changes, there may be ripple effects in our social systems. Psychologist Per Espen Stocknes says we need to "tell stories of the dream, not the nightmares." He suggests we might reframe the climate change message to emphasize opportunity, health, and quality of life. We should talk to people inside nonpolarized social networks such as sport teams, churches, neighborhoods, towns, and cities about the changes that might happen, are happening, to create better lives. These stories and changes might catch on. Savannah will follow Charleston, New Orleans will work with Norfolk, and finally, just maybe, Washington, D.C., will come around.

So that's where I headed. To find the stories of the people and places actually doing something about this spring that will be more sudden, unless we prepare. I wanted to see the kinds of information they have and how their expertise informs others, the partnerships they create, and their solutions.

Whenever possible, I talked to mayors' offices or city managers of the cities I visited. Often, I was deferred to someone lower on the pay scale, a sustainability manager or a deputy city manager who carries out the city's plan. These people are concerned with public safety and well-being. They have lived in their cities a long time and have seen changes unfold. They are practical, for the most part, and not ideological. In these places, adaptation to climate change happens in a nonpartisan, get-it-done manner. They take calls about fixing sidewalks, pot holes, the trash, and increasingly about flooding. That issue is no longer an ideological one for them but one disrupting their constituents' ability to get to school or the store. One of the first places I went was Norfolk, Virginia, where then mayor Paul Fraim said he was "jumping over" the issue as a political one. In the Hampton Roads region, home of the world's largest military to defend the seas, the sea level is not a conservative or liberal matter but a practical one, affecting the readiness of the base. The polarization of the issue that takes place on the national level is mostly absent when competent local government addresses real risks that people can sense all around them. There is no deciding which side you are on when the water rises around you.

I also talked to residents to learn what they see and what they know about the plans in particular places, and whether they think those plans are sufficient. I also spoke with experts in geology, ocean sciences, biology, conservation, city planning, and engineering to learn what they know of changes on the coasts, among plants and wildlife, and what potential solutions are required.

In each place, I asked about the changes occurring and the steps taken to address them, the short- and long-term plans. Though the problem is global in nature, it might have to be brought down to a local level before macrochange can occur. *Downscaling* is the term for bringing climate change models and statistics down to the regional or county perspective, as if zooming in on a digital map. One example of such a data output is a 2017 study published in the journal *Science* by a consortium of economists and scientists called the Climate Impact Lab. They examined the economic harm that could afflict southern places. The Northeast and West would fare relatively well, but parts of the South would be "hammered" said the study's lead author, Solomon Hsiang, a University of California economist. The study found that on the whole the nation's gross domestic product would decrease by 0.7 percent for every degree Fahrenheit increase in temperature (1.2 for a degree of Celsius), but southern places would be hit hardest because they are already hot and poor. The hotter temperatures would lead to a loss in productivity, increase in air conditioning and energy costs, and many heat-related deaths, if cities do not take preventive measures. These communities are finding out, or will soon find out, that the costs to prepare for change will be less than the costs to rebuild flood damaged infrastructure—not to mention the toll in human misery and displacement and to say nothing of the destruction of species and ecosystems, which the economists could not valuate.

One way or another, climate change will be a proving ground. We will either sink, in cases where the land is subsiding, or swim, finding ways to address these challenges. Those changes needed to adapt to climate change might be woven into other social and economic ones. We might turn the coming crisis into building stronger, more sustainable communities, as Naomi Klein argues in *This Changes Everything* (2015). For some, it will be difficult if not impossible to adapt to a dynamic system in a mode of accelerating change. Possibly, adaptation is a very expensive Band-Aid applied to a severe wound. But the people I talked to in the places I went do not see it that way. They are building capacity, resilience, until we stop adding fossil fuel to the fire that is heating up the planet.

When Rachel Carson penned her classic *Silent Spring* in 1962, the effects of DDT were noticeable. Birds were disappearing and human health was being affected. To mobilize action, she was able to draw on existing fears of the time about nuclear fallout. The noxious spray was not unlike radiation: invisible, deadly. And while the gasses that produce the heat-trapping blanket responsible for climate change are also invisible, they are natural, unlike DDT. Too, the effects are slower and more gradual, unlike a bomb. Carson intended her book to serve as a warning of what would happen if we did not change. Spring without birdsong would be silent. I see my purpose a little differently, as showing the change that will occur and how to cope with it. With adaptation, the effects of the storms and floods of spring (or other seasons) will be much less severe. In places like Norfolk, Miami, and Charleston, they experience higher spring tides, the large rise and fall of the tide at or soon after the new or the full moon. The *spring* refers to the springing forth of the tide during those lunar phases. Because they have nothing to do with the season, some have started calling them king tides. Spring tides happen twice a month, but king tides really happen a few times a year when the Earth, moon, and sun are aligned in a particular way, giving a boost in gravity from the moon's proximity to the Earth. A king tide is basically an especially high spring tide.

Though it can be hard to pinpoint particular storms as happening as a result of climate change, the overall trends are clear. And while temperatures and seas rise slowly, we have some immediate choices to make. If we act quickly and boldly, there is a small window of opportunity to prevent the worst. We can prepare for the changes by understanding what is happening and taking specific measures. There is "commitment" already in the climate change system. To minimize those effects will require another kind of commitment, the kind I aim to show in the coming pages.

Though I write about places in the South, the idea for this book may have come from a moment when I was far north. I had a chance to visit Glacier Bay in Alaska several years ago, and while the boat turned in a slow, mesmerizing 360, a large chunk of blue ice calved into the water below, crack-splashing into waves. Though such rifts can simply be a result of expanding glaciers, over the cries and camera clicks of the other passengers, the narrator from the National Park Service told of the evidence for climate change and of the efforts they were making in the park to cut emissions. She also told of how some places in the park system, like the one we were in, would transform in the coming years, altering the experience for future generations and how we would need to adapt to that change. I wanted to visit some of those most threatened areas before it was too late, like Glacier National Park in Montana, but I also realized that there are places in my own region susceptible to climate change and sea level rise and that some are trying their level best to confront it. Instead of melting glaciers, Cumberland Island National Seashore in Georgia faces erosion from rising seas, and the Everglades National Park in Florida confronts salt water intrusion and a loss of habitat.

I am not from the South, but I have lived in Virginia as long as I have lived anywhere. In my home state, I have hiked many of its trails, biked countless backroads, paddled and fished as many rivers as I could. Twenty years on, it feels like home. But I had yet to explore much beyond my state, including some major historic cities like Charleston, Savannah, Miami, and New Orleans. In part, the genesis of this book came out of an urge for going. I headed for the coasts because the issues related to climate change are most visible there and talked about. That's where I could find stories. Climate change will affect people farther inland, it will affect our forests and crops, displace species, but those changes are less visible, so far.

I include a chapter on West Virginia to highlight some of the changes happening in the interior and because, through the mining and burning of coal, they are implicated in the climatic effects of low-lying regions. Too, West Virginia must make an economic adaptation that other places will have to as well, as we transition from a fossil fuel economy, and they have experienced the wrenching trauma of both climate change and economic transition. That trauma and its attendant psychological dilemmas are features of adaptation, not on the infrastructure side but a cultural adjustment that will become more prevalent as the climate shifts. The highlands, too, could be a place where those in low-lying areas will migrate to as the coasts become inundated and unlivable. And yet another reason is that in addition to fossil fuels, West Virginia possesses great reserves of natural beauty that nourish in a time of upheaval.

"The sense of place is deep within us," one geologist said to me, which might be a reason both why we notice small changes and why it becomes difficult to fully accept their meaning. We want to stay in a place we have gotten used to, despite the projections. But we seem to assume a particular lay of land or shoreline will remain fixed and constant, that trees will stay trees. The reasons for doing little about climate change, or denying it altogether, are varied and complex. Misinformation plays a part, as does motivated reasoning or bias. But some I talked to were simply more alarmed by immediate threats, such as terrorism or crime, than something far off. In the North Carolina chapter, I write about how the brain has difficulty taking in the temporal and spatial data. In the Georgia section, I discuss our optimism bias, the belief that things will turn out well, causing us to put our faith in technology or other powers. It may also be that we simply do not live long enough to process the big picture, though some project we are arriving at the point where we will notice changes. And we have sometimes tried to address one problem, such as damming and diverting the Mississippi River to control flooding, but have created more, a lack of sediment downstream. It is like we refight the last battle but are unable to envision the coming war.

I was preparing a final version of this book just as Hurricane Harvey (2017) stalled over Houston and prepared to dump historic level rains. I address that storm in a postscript to the Louisiana/Texas chapter and include some updates from Irma in the South Carolina and Florida chapters as well. By the time the book is released, there will likely be other catastrophes to strike these places, new information, revised assessments or projections. The information I gather here should help put those new threats in context.

The sea is rising everywhere but not at the same rate in all places. Some places have worse subsidence, or gradual sinking of land. Each place I visited has a different set of circumstances, politics, and civic culture. Norfolk has a navy to defend it, but in a city that used to export the coal responsible for much of the global warming, they are also sinking. Charleston has experienced three major flood events in successive falls. They are digging massive tunnels underground to help deal with their flooding. Miami is not sinking but sits on porous limestone, and New Orleans has sunk so much it is below sea level, requiring levees and pumping systems. Houston and Galveston have valuable and vulnerable refineries to protect, along with a growing population. Each has decisions to make about where and how to spend money, if it can be found. And if found, will it alone be enough to deal with the problems they will experience? The alternatives will not be easy, as they will be faced with "strategic" or planned retreat. Eventually, in the not too distant future, some places will have to be let go.

Gradually, I began to see connections among them like dots in a pointillist painting or one of those pictures you step away from and the image within becomes clear. We each play a role in climate change and the storms it causes, but they affect rich and poor in different ways. Wealthy places can fortify, spend money on protection. Poorer places do not have the luxury of such spending, with more immediate and pressing needs. And poorer people will have a harder time relocating, or evacuating, and will bear a disproportionate brunt. But when it comes to climate change, we are all in the same rising, indefinite boat.

Climate change touches everything, a story and calamity unlike anything else. It's flooding and drought, bigger deserts and shrinking forests, melting at the poles and tropical disease moving toward them, and a very uncertain fate for most species on the planet. I have tried to seek out the stories of people trying to grasp these changes, people who attempt to understand both the past and the foreseeable future as well as the systems of survival.

Several persons I asked to interview wondered why an English professor was interested in climate change. Of course, writers of all stripes—poets, journalists, philosophers, star gazers—have sought to comprehend the natural world. Recently, novelists, such as Margaret Atwood (*Oryx and Crake*, *The Year of the Flood*), Cormac McCarthy (*The Road*), Ian McEwan (*Solar*), and Barbara Kingsolver (*Flight Behavior)*, have turned their attention to climate change as it is not just a scientific phenomenon but also a cultural one leading to displacement, disruption, and existential challenges. As I teacher,

I talk with students about the complexities of these and other texts and the subjects within. As a writer, I have always been fascinated by how humans value (or devalue) the landscapes they inhabit and by the stories we tell about our past and those we imagine for the future. As a citizen, I simply wanted to know more about what is going on, particularly the fate of life on Earth, and why the issue of climate change has failed to take hold in certain corners but has moved to center stage in others.

While each place has a different circumstance, a theme began to emerge as I drove to them: the fleeting nature of things, their transience. I live on the western side of the Blue Ridge Mountains, and I travel up and over them to get to the coast. During one of these journeys, the clouds stalled over the ridge and they looked like a shadow of their former much higher selves as the rain washed highland silt toward coast. On the beach, I stood on that pulverized rock and rubble from long ago, which was itself shifting and wearing away.

We struggle to understand these changes, to confront the forces that would eradicate us, to leave a lasting mark. Those efforts would seem to take the visible form of a levee, seawall, or raised structure. But all around is evidence of the damage, such as fallen tombstones, and a sense of foreboding, like we are running a race between destruction and salvation. Even in these places so conscious of a rich historical past, little remains the same. Not even these old mountains. If trends continue, we may have little to show for the changes but the stories of these places, their efforts to adapt and endure. In seeking out these stories, I hope to bring these beautiful dying places to life.

2

Our Best Defense

Shelling the Naval Base, Virginia

Our story might as well begin thirty-five million years ago when a fiery meteor slammed into the mouth of the Chesapeake Bay, creating an impact so violent it likely killed everything on the East Coast. Then, the Virginia shoreline was somewhere near Richmond. Tropical rain forests covered the slopes of the Appalachian Mountains. The bolide sent tsunami waves crashing into the Blue Ridge and created an imprint that determined the eventual location of Chesapeake Bay. It weakened the sediment across a fifty-mile area. Norfolk is at the edge of that zone, and some scientists think the ancient cataclysm may be one reason it is sinking.

I was headed to Norfolk after leaving class one afternoon when it rained in buckets. Heavy thick sheets pelted the windshield. Just before my entrance to the highway, tractor trailers had collided, jackknifed, cutting off the northbound flow. On the opposing side, farther north, something similar had happened. As I crept on to the normally busy interstate, rain letting up, I had both lanes weirdly to myself and no one was coming in the other direction. I had just finished teaching Cormac McCarthy's *The Road* where something like a meteor struck and the characters make their way slowly to the coast, their possessions loaded in a shopping cart. As in the book, the scene was unnervingly apocalyptic.

Norfolk is due east of my house, but the easiest way there is to drive north on the highway to where it cuts through the mountains and then southeast through Richmond. I was to meet Capt. Chris Moore of the Chesapeake Bay Foundation at the Haven Creek boat ramp in the morning.

Though Norfolk is three hundred miles from where I live, the same storm had affected the coast. Easterly winds had pushed water up into the streets, blocking my path. Behind a McDonald's, where Llewellyn Avenue dips, pooled a foot of brackish water. My GPS would not reroute. It could show traffic but not water, only it would someday, according to one person I would meet.

The water was so high Moore said he had difficulty launching the boat in the water. Haven Creek is a creek no more, filled in, but when there is flooding, that is where the water wants to go.

Moore is in his early forties but exudes a boyish enthusiasm found among boaters. We were joined by Thomas Quattlebaum, a younger associate and the Chesapeake Bay Foundation's sea level rise expert. Moore wore water boots, dressed for the flooding. Quattlebaum, the sea level rise expert, forgot his—he was not expecting the flood. Nor would anyone really. It was a bright sunny day.

Moore backed the *Bay Oyster*, a white skiff, into the Lafayette River and piloted us forward. From the boat, we had a view of the Lafayette Towers apartment high rise, water near the front door. The men noted a storm water drain coming off a nearby parking lot. To those protecting the health of the bay's water quality, the drain was already a problem as it brings the runoff of oil and debris. But with the flood, the opening was close to being buried under water so it would back up and make the flooding worse.

Around the corner, we toured some of the flooding. The water was up over someone's lawn chair, topping a nearby wall, a foot of water in the streets. A few snowy egrets plucked fish out of suburban lawns in an area called Colonial Place. On Mayflower Avenue, cars were parked high up on the sloping driveways, but they would not be leaving soon. Not unless they wanted to hop in our boat.

One red truck was out inspecting something, and Moore steered the boat over in that direction. Near the corner of Mayflower Road and New Hampshire Avenue, Justin Haberstam had driven through the flooded streets to double check an $8,000 gauge that runs on solar power. The gauge read from a stilling well that dampened the wave action, helped the graph "lose the bounce," allowing hydrologists to show flooding trends. This information was sent digitally to the city so they could communicate with residents about street flooding. Ultimately, this information might be available to a phone's app so residents would not be stranded behind the McDonald's.

Moore knifed the boat through the choppy waters. He showed me how the concrete piers of one particular yacht club had to be raised because the boats could become unmoored, float away, with rising seas. In the distance, we could see the Norfolk International Port and the navy base, the largest on the globe.

Norfolk exists because of water. Bounded by the Chesapeake Bay and the Elizabeth and Lafayette Rivers, sliced by coastal creeks, Norfolk has always experienced some form of flooding. The city built ships during the Revolutionary War. The port is still a terminus for the Norfolk and Southern Railroad. It travels the route I did the previous day or a more direct one, bringing coal from the mountains farther to the west. Ironically, the burning of coal has been responsible for as much as 25 percent of greenhouse gas emissions, which caused heat to be trapped in the atmosphere and sea levels to expand and rise—a preview of which we were seeing.

On the day I visited, some of the naval base piers had to be shut down because of the flooding. It wasn't the ships—they could float. But the infrastructure that supports them was vulnerable. Supporting the piers are thick cables and pipes for fresh water, sewage, and electricity attached to the underside, open to the ocean below. The flooding of housing and maintenance facilities is an issue, as is access and egress. Imagine a scenario where sailors have to get to ships. The enemy is upon us. Only there is too much water.

On a bus tour of the facility I learned that the naval base provides $50 billion in revenue to the area, employing eighty thousand active duty. The base sits on more than six thousand acres in Norfolk, on a point of land near where the Chesapeake Bay meets the Atlantic Ocean. The tour guide, herself a midshipman, pointed out how some structures had been updated for higher hurricane ratings. At the end of the tour, I asked her if flooding on the base was a problem. She said it was mainly off base, pointing to a wrought iron fence perimeter that would seem to provide little in the way of flood protection.

While Moore drove to our next spot, Quattlebaum spoke about the urgency of the situation. "Every day of inaction the risk grows," Quattlebaum said, as we spied a pod of disinterested dolphins off the bow. Since 1927, the National Oceanic and Atmospheric Administration has had a tidal gauge at Sewell's Point, the peninsula where the naval base is located. It has recorded the highest relative sea level rise on the East Coast, nearly fifteen inches, just below the knee of an average adult. And that rate will accelerate in the coming years. Michelle Covi, a member of the Mitigation and Adaptation Research Institute at nearby Old Dominion University, told me the city used to experience two flooding days per year. Now they get ten to twenty. Because of the flooding, a new moniker has emerged for the Hampton Roads region (Norfolk, Virginia Beach, Newport News, Portsmouth, Suffolk, among others): "Dampton Roads." The area is also known as the Tidewater because of the effects of tides on local rivers. According to Steve Nash, author of *Virginia Climate Fever* (2014), three-quarters of Virginians live within twenty miles of the ocean, bay, or a tidal river.

On my first visit to Norfolk, I talked with Ron Williams, then deputy city manager of Norfolk (now at Virginia Beach). Williams kept referring to "recurrent coastal flooding," "nuisance flooding," rather than sea level rise. "We're not wrapped up in the politics of causality," he told me.

Williams referred me to a copy of their Coastal Resilience Strategy created in 2012. It plans both for the immediate future and for 2100. By the end of the century, they could be looking at three feet of sea level rise, but that is a low scenario. Scientists at the Virginia Institute of Marine Science (VIMS) say they could see more like five to six feet on the high end. Williams kept referring to a map on the wall that highlighted some of low-lying and legacy (or former) creeks. There are some $1 billion in improvements listed in the plan, and Williams wanted to emphasize that they had a plan. He also wanted to emphasize resiliency, and Norfolk is one the Rockefeller Foundation's 100 Resilient Cities, a global network of cities including Norfolk and New Orleans dedicated to urban resilience. Each city has a chief resilience officer who coordinates these efforts. Said Williams, "Norfolk was bombed by the British, bombarded by the Yankees, and survived two yellow fevers. We'll bounce back."

Quattlebaum could see a "last man standing" scenario, where some houses would be declared a repetitive loss because of repeat flooding. Once one goes, the others in the neighborhood could divest. "But people want to live by the water." Quattlebaum himself did. He was from South Carolina and spent a good deal of time riding a surf board. His belt buckle was a picture of a wave. "We're facing a significant challenge in our lifetime, our children's lifetime, to make these places more resilient for the future." At one harbor Moore steered the boat into, water was over a dock, up to the bottoms of gas pumps. On the other side, in a residential neighborhood where they had hardened the shoreline with concrete, some of that had degraded. The road buckled some, probably from flooding. The base of a few live oak trees soaked in the water.

Quattlebaum said they may be facing a scenario where businesses will stop coming to town unless the city shows it is doing something—that can show they are serious. The city lost a big brewery to Richmond. Norfolk has been working on flood planning with the Dutch, but in the Netherlands, everyone is impacted. Not everyone in Richmond or Charlottesville is, which has made funding a challenge.

It frustrated him that people could not see "past the end of their noses." "Each disaster wakes people up," but it was disconcerting to see such inaction, putting things at risk.

Quattlebaum tries to speak to communities about sea level rise, presenting it as an opportunity so as not to scare anyone. But he heard the city manager of one Hampton Roads community say "we are not fucking retreating." He was driving around town with a city councilperson of another, Poquoson, who told Quattlebaum that he had his house raised using FEMA money. Quattlebaum said that some communities are using that same money to acquire flood-prone properties and help people relocate rather than raise. The councilperson told Quattlebaum that he would tell people who suggested such an idea to "go to hell." Furthermore, he said that anyone who suggested retreat would be voted out of office. What if a flood comes and you need to get to the hospital? "We'll get in our boat and take ourselves to hospital." One community was purchasing a hydro-boat with a nozzle for fire, though first responders could have trouble getting to it if there was a flood.

Idling on, Moore pointed to an electrical unit that looked precarious. Nearby was a pump, able to withstand a flood, "silenced by Dri-prime." We saw an osprey dive and a cormorant nest on piling number 13.

Their answer to the problems we saw was to look for green infrastructure opportunities at every chance. The Chesapeake Bay Foundation would restore wetlands or find ways to capture storm water so that it did not run directly into the rivers. And rather than harden the shoreline as if with armor, both Moore and Quattlebaum preferred a "living" shoreline with natural habitat, such as a wetland or grass buffer with oyster reef. The hardened structure can worsen erosion, while the living shoreline traps sand

and silt, straining it from the wave action, building up the coast. And a living shoreline further provided the benefit of improved water quality and a healthier estuary.

Moore steered over near the Hermitage, an Arts and Crafts style building once owned by the Sloane family who made underwear and socks for the navy. It was now a museum and a good example of a hybrid living shoreline, where some structure supports the vegetation. The original brick wall and historic garden was still intact behind a buffer of marsh grass and foliage, but Moore motioned to a point that no longer was.

On an earlier visit to Norfolk, I went to the Chrysler Art Museum, named for the automotive heir whose wife was from Norfolk. It is located on one of these creek inlets, now a crescent-shaped seawall, prone to flooding. The area, called the Hague, was settled by some Dutchmen (who named it for their capital), who are now invited back to consult. The area is older with cobblestone streets and stately homes. The Unitarian Church was for sale in part because of the flooding. One of the docents told me the art museum sometimes gets questions from those who would lend priceless artifacts: These are not in any flood danger, are they?

Across the way from the Hermitage is the Virginia Zoo, another site of a Chesapeake Bay Foundation coastal project. They cleared some trash, and some of the sand went to the tiger exhibit. There was an oyster reef "working like it should," meaning lessening wave activity and controlling erosion, along with spartina grass and wax myrtles, keeping in the shore, filtering the water.

Groups like the Chesapeake Bay Foundation work to adapt to changing conditions one area at a time. I also met with Skip Stiles, director of Wetlands Watch, a Norfolk-based advocacy group dedicated to protecting shoreline habitats. Stiles, in his sixties, realized that though saving the marsh was good for coastal protection, nothing would help if there was no coast. He once worked in national politics as chief of staff to the late California Democratic representative George Brown, who in 1978 launched the first federal climate change research program. But Stiles began to realize he could best effect change on the local level.

When he looked for existing solutions to help communities cope with sea level rise, he could find nothing that was "off the shelf." The community in question is Chesterfield Heights in Norfolk, historically black and disadvantaged economically. The average income in the area is $14,000. Most have lived in their homes for generations. They own them and they are paid off. They neighborhood was built around 1910 on the trolley line, and several of the homes are on the National Register of Historic Places. During Hurricane Irene (2011), residents were cut off from going anywhere. "I couldn't get past the curb," said one. The neighborhood was built near one of the legacy creeks that were filled in or developed. Said Stiles, "a creek wants to be a creek"—it is where the water wants to go when it floods.

With money from Sea Grant, a National Oceanic and Atmospheric Administration (NOAA) program devoted to coastal research, Stiles put together a kind of design competition involving area schools, engineering students from Old Dominion University, and architecture students from Hampton University. At the presentation of these designs, Troy Hartley of Sea Grant said "there is a gap in the tool-kit for adaption." And Stiles said he hoped the solutions would be nature based rather than "just pouring concrete" for seawalls.

The students presented their solutions before a city-wide Watershed Task Force at Stover Library. The purpose: how to keep a community safe from the flooding it experiences now and would experience in the future.

The students came up with at least five solutions. One involved the living shoreline I was seeing on the Lafayette, only the Elizabeth sees even more erosion from barges traveling up and down the river. They would also replace street bricks with pervious pavers to let the water drain. Three of the solutions pertained to holding the storm water, at least temporarily. They created bioretention areas, landscaping to absorb some of the runoff, therefore improving water quality. I sat next to their professor, Mason Andrews of Hampton, who told me residents get a good deal of their protein from the river, which is impaired.

They also proposed something like a basement cistern. The residents' basements flood, so some have thought of filling them in, but this will only displace the water, making flooding worse. The students discussed basement cisterns to hold the water until the storms subside and the water can be pumped out. Finally, they had designs for something like a yard cistern. Right now, the water from the roofs runs off and through a gutter and into streets. What if they could capture that water? The students developed plans for a cistern system that would slope away from the house. They would be like planters, with grass or plants on top, but underneath there would be a capacity to hold water. These would take pressure off existing storm water systems.

One obvious solution would be to raise the houses, but they are historic and the residents like sitting on porches. Raising them (the current building code calls for the first floor to be three feet above predicted flood level or freeboard) would change the character of the home and community. The vegetative cisterns would slope away from the porch, adding character, not too different from a lushly planted garden.

The students, including Zach Rodriguez from Hampton and Alex Carlson from Old Dominion, see sea level rise as an urban planning problem. They ran a model and saw a 90 percent reduction in flooding. The cost? About $1 million.

For Stiles, an economy could be created, selling their expertise to places like Savannah and Charleston (with historic character) that will also need to adapt. He said the Dutch get 4 percent of their GDP from consulting on sea level rise.

At city hall, Williams talked of an economy of water management. "If you have an idea, come try it out in Norfolk." But he said the Chesterfield Heights project is the "tip of the iceberg," an apt metaphor when talking sea level rise. Still, often these solutions are "done to" rather than with the communities. The scary words they use in places like Norfolk are "intentional departure" and "strategic retreat." But maybe not yet. Maybe they can lead by finding ways to adapt before such measures are necessary. Maybe they can also rise above the political debate on climate change for the good of their community.

But geologically, they are sinking. Sea level rise is relative. What one area experiences is not the same as other places around the globe. On average, places are experiencing a little over 3 millimeters per year of rise. Norfolk is experiencing more like 4.5. The area is seeing a rate of relative sea level rise twice the global mean average due to geological factors relating to subsidence, including the rebound from a former glacier off the coast. It once pushed down on land out at sea, in effect raising the coast as one who sits on an air mattress raises someone on the other side. But if they leave, that person comes back down. The effects of the meteor/bolide are another factor, as is groundwater extraction.

One reason the land could be sinking is the result of households and businesses sucking it dry. Geologists have found a "cone of depression," where the water table is lowered faster than it can be refilled, over a paper mill near Franklin, Virginia. Such drawdowns can exacerbate subsidence. Using an extensometer, a long rod placed into rock layer to measure ground surface relative to bedrock, geologists found that between 2002 and 2015, as the paper mill reduced water and eventually closed in 2010, the ground level rebounded thirty-two millimeters.

Ted Henifin, general manager at the Hampton Roads Sanitation District (HRSD), had the idea to replenish the aquifer by injecting water from its treatment plants into a huge aquifer deep beneath the surface. He hopes it will slow and eventually reverse the sinking of land throughout the region.

SWIFT, Sustainable Water Initiative for Tomorrow, could pump as much as one hundred million gallons a day into the Potomac aquifer, all of it treated to drinking-water-quality standards. Henifin said HRSD likely would pursue the project even if it does not help with subsidence. Recharging the aquifer would benefit groundwater users by keeping seawater out, and HRSD's discharges of treated, but still polluted, water into the Chesapeake Bay and its tributaries would all but cease. Instead, water, treated to standards higher than normal discharges would be pumped into the aquifer at low pressure. According to Henifin, "it will be the biggest aquifer recharge in the country."

Will it show what many folks who support HRSD's plan are hoping: that we can slow the threat of the sea coming at us by rehydrating and reinflating the ground deep beneath us, like putting water back in a sponge? Henifin and the HRSD are betting $1 billion, the project's cost, that the answer is yes. "It won't stop sea level rise," Henifin told me, "but we think it will provide some near term relief."

Just before lunch, the crew trailered the boat and drove over to the Brock Environmental Center, built to the highest environmental standards and to house Chesapeake Bay Foundation's Hampton Roads staff and a local conservation group, Lynnhaven River NOW. Before I had a tour of the building, Chris drove across the inlet to a dockside restaurant. On the way, he briefly nosed the boat out into the bay chop to show me some huge sand renourishment projects. The waves picked up, a succession of them rolling the small skiff. I was glad it was before lunch.

Earlier in the summer, I made a trip farther out in the bay to Tangier Island, a dot of mud and marsh in the bay divided by smaller marshes and tidal flats and accessible only by boat. I took the ferry from Onancock on the eastern shore of Virginia, fifteen miles to the island, a forty-five-minute ride. Out in the bay, the boat pitched fore and aft, kicking up spray onto the windshield. With my legs spread out to absorb the motion, I asked Capt. Mark Crockett, who was born on Tangier, and shares a last name with the founding family, what he knew about the flooding.

"We're not going to wait around anymore. We're going to do something about it," he said.

"Like what?" I asked.

"Sink barges if we have to."

That night, I had camped at Kiptopeke State Park at the peninsula's southern point and site of the ferry before the Chesapeake Bay Tunnel was built. A few hundred yards out into the bay loom several massive, decrepit concrete ships, partially sunk, arranged in neat lines to form a breakwater. They once housed seamen. Now, a thriving ecosystem. Captain Mark had something like this in mind, to buffer waves and protect the shore. Islanders were set to do this but were called off, in part because of an environmental review. From what I could gather, the Chesapeake Bay Foundation had a role in it. They had concerns about the hazards of sinking an old vessel.

To learn about existing plans for the island, I met with David Schulte, a marine biologist with the Corps of Engineers in Norfolk. Their building sits on a point next to the Elizabeth River so they have a good view of sea level rise. I had to be cleared by security at a front gate and inside the building before David and I could meet in a conference room. He wore a business suit and sipped from a glass cup of loose-leafed tea that said "half empty." "It's hard to be optimistic when you study this stuff."

Schulte authored a 2015 paper on climate change and the impacts on Tangier. In it, he details the rates of sea level rise and subsidence. Tangier Island and nearby Norfolk are in a "hot spot" of sea level rise.

In the largest tidal estuary in the nation, Tangier is also home to the best soft-shell crab fishery anywhere. When I visited, the day's catch was off to Baltimore, New York, even Pike's Place Market in Seattle.

Schulte estimates "ecological services," which are words scientists use for stuff we eat or drink, in the millions annually. Schulte would also say there is intrinsic value in the living organisms themselves. It is just harder to put that in a scientific or army report.

By digitizing the coast, and comparing it with maps since the 1850s, Schulte and his group found that as of 2013, only a third of the Tangier Island's land mass remained. The 770-acre island was losing 8.5 acres per year. His team looked at historic records to separate out the impact of major storms and determined that relative sea level rise is the "driving factor for land losses." Under a midrange sea level rise scenario, it would lose more in the next fifty years, "and the town will likely need to be abandoned." He submitted the paper with all the data referenced, figures checked, the economic losses calculated, but the editor wrote back. "Great. But you need a conclusion." He added one. "The Tangier islands and the town are running out of time, and if no action is taken, the citizens of Tangier may become among the first climate change refugees in the USA."

Schulte outlined a rough engineering plan costing around $30 million that involved segmented breakwaters (like rock jetties), new vegetation to restore wetlands, and pumped-in sand to restore beaches and dunes. For the latter, they spray sand much like some landscapers spray mulch. He said some structures would have to be abandoned, some elevated. But what he not could design is how to pay for it. Congress would have to step in, or the state. Tangier themselves could not pay for it as other cities can. I asked if they had a lobbyist or something, but Schulte said "they don't really have anyone advocating for them."

For Schulte, it was "worth saving the island just for the ecological value alone." And how do you put a value on one of the last island cultures of the bay? The island was placed on the National Register of Historic Places in 2014, and the community has been there for hundreds of years. John Smith visited the island in 1608, and it was settled in the eighteenth century though early land patents date back to 1670. Many residents came from West Country England, Cornwall, and because of their isolation, many have retained a dialect of that region, even in the age of TV and Internet, one that has drawn the attention of linguists. "There" has a hard "d" as in dare, "out" and "about" sounds Canadian, more like "boat." It's a brogue of clipped consonants and unusual cadences.

Yet, as much as a physical deterioration, there has been a social erosion too. Most of the islands residents are watermen, up early for crabs and oysters, or they work on tug boats. But young people are leaving the island in part because of regulations against overharvesting. The Virginia Marine

Resources Commission put a cap on commercial fishing licenses. Young people that attend college rarely return. According to the 2010 census, the population was 727, but most people I spoke with put that figure at closer to 470. Still, Schulte thought that relocating would be very expensive. It would be more cost effective to keep people in place and invest in protection. Shielding the island could be cost-effective as well as the right thing to do.

Carol Pruitt-Moore agreed to take me on a tour of the Uppards Island, one of the Tangier Islands. Residents of Tangier used to say they were going "upward," hence the name. It's a small island at the northern edge of Tangier, once connected to it, but now water-soaked and separated by the channel used by waterman and the ferry. On the way out to the island, Pruitt-Moore waved at her husband, returning from tending his crab traps, little to show for the morning's effort. There were several abandoned shanties where we boarded a boat. Pruitt-Moore said tourists ask them if that is where they keep their cars. There are, in fact, very few on the island. I brought my bicycle to get around.

Her boat was named *To Oz*. "Because I never know where I'm going. And I don't know if I should smile or cry."

She beached the boat in a small harbor made by erosion. We stepped off and began walking the island, site of a former settlement once known as Canaan. Pruitt-Moore is in her late fifties, tall with short blonde hair, and a seventh-generation islander. She makes the trip almost daily. As we walked, she favored one leg. She told me the island was abandoned in the 1920s. Walking the beach, we saw remnants of a former hunting lodge. Her uncle used to run it.

We walked among a former shell midden, oyster shells everywhere, crunching underfoot while flies buzzed us and the wind blustered. She frequently comes up there to hunt for artifacts and arrowheads. They found coins from the 1800s. "My friend Stacey found 94 arrowheads in the month of May." The Pocomoke used to travel through here, at least in summer. One anthropologist told her the heads came from as far away as Kentucky and Arizona, through trading. The British established Fort Albion here during the War of 1812, launching raids up and down the bay, particularly on Baltimore's Fort McHenry. One Francis Scott Key was detained on one of these ships, and while he waited for the battle's outcome, he composed a poem, an anthem even. Twelve thousand troops once occupied the island, and nearly a thousand former slaves took their first free steps on Tangier's beaches, though they were later sent to British colonies after the war. When John Smith saw it in 1608, he called the complex of islands Russell's Isles, after Dr. Walter Russell, the expedition's "Doctor of Physic."

"This is where people lived. The graveyard was here." A few gravestones stood at odd angles. Some lay flat on the ground, knocked over by waves, contents sometimes spilled out. One of them had the name Pruitt, 1836–1901, a distant relative she said. There were remnants of a well, now filled with salt water at the island's edge, an axle off a trailer or truck, tires, and a bathroom sink. We found a white handle from a ceramic cup, likely from the steamboat that used to come through.

And then I nearly stepped on a human bone. "That's a leg," she said, as if a common occurrence. "Don't step on that." I wasn't about to. Not now that I knew what it was.

She pointed to a spot on the island that had disappeared, her voice trembling a little. "Right here was land. From that point to that point, that was land last year." About the ground we were standing on, "this will be gone next winter." Waving behind her, at the grass and shrubs, she said "if we lose this, we're in trouble."

She motioned toward a wave breaking into shore. "That was sand when I was a kid. Lush green grass with fig trees and roses and wild asparagus. Goats and chickens roamed over here. I used to walk this with my father." She stopped. "This was land last year," she said again in disbelief, pointing to something that was not there. A preview of what was to come. The bay was mostly calm now, little foam or froth, but it has been biting away at the land for centuries, and could turn nasty in no time.

"It makes me sick, she said," looking off into the wide expanse of the bay. "To think we're some kind of test project. We've been studied. We're studied to death." She mentioned some of the work residents did, writing letters, meeting with officials. "We're just not important enough."

She spoke of the wildlife she sees, a further reason to save the island. Turtles. Sting rays. Dolphin pods by the thousands. One particular day, striped bass were so abundant, "they were flipping themselves on the shore." She described how she got down on the ground, "I don't know why I'm telling you this," and caught one with her bare hands but was "chicken" so she let it go. Oyster catchers and pelicans. Her voice grew livelier now, her face glowing. She came up here one day when monarch butterflies coated the western shore, like it was on fire. "The whole beach an orange blanket." But Hurricane Sandy took away the goldenrod. And they found five complete skeletons from the graveyard washed out of coffins. Archaeologists with the Virginia Department of Historic Resources retrieved the remains. Another storm, she worried they'd find more.

On the boat ride back to the harbor, Moore-Pruitt told me she was not a believer in human-made climate change but would support whatever angle they could use to secure a wall. We stopped at her house before we boarded the boat because she had to leave something with her grandkids. Two flags flew high: one from the U.S. Marines, and another the flag of Israel, which some conservative and evangelical Christians fly in solidarity with the holy land. The mayor, James "Ooker" Eskridge, flies one too.

I asked about the state of things now. In 2012, Governor McDonnell came. They signed some papers. Construction was supposed to begin in 2014, but now it seemed they were awaiting further studies. "Studies have been done and done and done." "Maybe we'll invite Donald Trump," she said. "Maybe he will help us." I wasn't sure exactly what she meant and didn't ask. I liked Carol, her feisty personality and love of nature, family, and her island community. But I was thinking it: *You mean build a wall?* So I joked, "Would you want to put up with a casino hotel on Tangier?"

She laughed. "Whatever works, right?"

The heat index for that day was 115°. I forgot my hat but Pruitt-Moore gave me one from the Chesapeake Bay Foundation, where her husband used to work, but they had some kind of falling out, perhaps disagreeing about restrictions on catches or the sinking of ships. "You can keep the hat." I had noticed on the way into the harbor that Port Isobel, an island to the eastern shore of Tangier where the Chesapeake Bay Foundation has an educational center, had about six protective breakwaters, but theirs were built using donations.

Renee Tyler, the town manager, might have gotten the idea from them. She began a campaign on generosity.com to raise money for the seawall. "We are in DESPERATE need of seawall protection around our entire island as well as restoration projects to rebuild our island to the size it was in the 1700s." So far, she had raised $1,400 out of a $2.5 million goal.

After leaving the harbor and thanking Pruitt-Moore, I found Tyler in her office at the town hall. It's a small prefab building out by the airport on the southwestern side of the island, mostly one room and some closest space, faux wood paneling, an air conditioner spitting out cool air. I had hoped to find out whose hands the seawall project was in, where it was stalled, who I could follow up with. But she did not know and was frustrated.

Tyler took the job as city manager first as a temporary gig, but they kept her on for ten years. She said she wasn't a politician, hated small talk, "what you see is what you get." She wasn't a scientist either so was learning about the dynamics of climate change. Tyler was not real forthcoming with her answers, a bit terse, but finally opened up when I asked if there was a plan B, of moving off the island. She said she still had faith. "It's our land. Our culture. Our way of life." "It's not my house," she said. "If it burned down, I'd miss it. But this is where I want to stay. It's my community. When there's a birth of a baby, a funeral, someone's sick—everyone comes to together."

Earlier that summer, there was a *New York Times* article about the fate of Tangier. She was still reeling about some of the comments. Somebody by the handle Tournachonadar wrote, "Do not, repeat, do not use my hard-earned tax dollars to bail out a few people on an insignificant spit of sand in the middle of the Chesapeake Bay." Jack wrote "now is the time to spend the money to help some folks move higher or inland, rather than trying and failing to save low-lying areas destined to wash away over the years."

Tyler received two more donations to generosity.com after the article. They were selling T-shirts that said "I refuse to be a climate change refugee." More to raise awareness than money, she said. "We need more than T-shirts. Good grief."

She spoke of the seawall out by the airport. They were losing twenty to twenty-five feet of shoreline per year until it was created. Now they lose nothing. "So it works."

I asked her to make her plea to the public, state the case. She said, like Pruitt-Moore, that it was time for action. She thought the studies focused too much on the environment. "They see the environmental side more than the human side." She thought experts worried too much about encroaching on the marsh, but without some protection, the marsh was going to disappear anyway. "It makes no sense."

She tried to find in her email the name of the person who stalled the state portion of the money in appropriations. While she was looking, a couple came in asking about real estate. They wanted to find out about vacant properties. There are about sixty of them on the island. Tyler had no list of the owners on hand. They were vacationing and thought they would look for themselves, having found only six listings online. When I checked, there were three. A single-family home, two thousand square feet, was going for $90,000. The historic Hilda Crockett Bed and Breakfast, five thousand square feet, was going for $300,000. The couple had looked in on the Sunset Inn, but it was in bad shape, run down and abandoned. I had seen it on my bike ride. It probably experienced some flooding. Tyler had the name of that owner.

I left them to do business and rode my bike around the island to explore. I pedaled past the Tangier Oyster Company, a new venture Tyler hoped would bring jobs. The day before I arrived, the governor's wife, Dorothy McAuliffe, had been to Tangier to promote the Virginia Oyster Trail, in hopes of drawing tourists eager to eat through the tidal region.

I rode out across a bridge over one of the tidal flats into an area called Canton, believed to be the area first settled. I rode streets until it seemed I came out to a house closest to the bay. Joan Thomas and her husband were working on their transportation, a souped-up golf cart, raised on planks for storage so it would not flood. Thomas is the surname of the Methodist minister who was known as the "parson of the islands" in the 1800s. When the British headed for Baltimore in 1814 (during the War of 1812), he prophesied defeat for the redcoats.

Thomas invited me to walk her property. The grass, neatly mowed, had turned to something more like marsh grass. It crunched underfoot, and there was some black algae or seaweed in the soil, brought in by the sea. They had built a small berm in the back, but a few fiddler crabs poked in and out of burrows. And there were hay bales also in use for keeping water out. She had been there for forty years and seen things get worse and worse. She remembered walking to a candy store, in what felt like a mile to the east. Now, we were looking at blue water past green shrubs, silver breakers some three hundred yards away. An osprey flew by, headed for the sea beyond the shrubs and grasses.

I asked the couple what they thought of their prospects. Like Tyler, they were hopeful for a solution. Mr. Thomas had high hopes for a new representative, Republican Scott Taylor, a former Navy SEAL. Unlike most in his party, Taylor has said that sea level rise is a reality in his district, which includes Tangier and Hampton Roads. "Data shows that there's a different sea level you know over the past 90 years. So I certainly acknowledge that it is an issue and it certainly is for us in that area," Taylor told a radio reporter.

Virginia has come a long way on the science of climate change. The state once made national news regarding climate change and its politicians. In 2010, Virginia attorney general Ken Cuccinelli asked the University of Virginia for material related to Michael E. Mann, a leading climate scientist who was an assistant professor at the university from 1999 to 2005. Cuccinelli alleged that Mann had defrauded taxpayers by manipulating data. A judge found insufficient cause, and the suit was widely condemned as an assault on academic freedom and an attempt to silence Mann.

And Virginia once had a state climatologist, Patrick Michaels, who Mann has described as a "climate change contrarian." He has contended that climate change will be minor and may even be beneficial. But his funding was coming from sources with a vested interest in him saying that, including the Western Fuels Association, an association of coal-burning utility companies. Science historian Naomi Orestes has confirmed that such groups took a page from the tobacco industry as "merchants of doubt," recruiting sciency spokespersons to make dubious claims about "uncertainty."

Representative Taylor has opposed cuts to climate science, especially to NOAA. The scientific agency is under the Department of Commerce. Its core mission of forecasting extreme weather was once relevant to protecting property and business. Some have suggested it should be moved to the Department of Defense, where the Corps of Engineers is, because it helps defend us. Then its funding would be less subjected to the whims of politics.

I reached out to Taylor's legislative director, Reginald Darby. He was aware of the land loss on Tangier. He said seawalls have been "obstructed by a lack of funding and burdensome regulation," by government inaction, and that the congressman supported a plan to "offer immediate relief from erosion at no cost to taxpayers." He wanted to sink the out of commission barges, donated by a salvage company.

Susan Connor, chief of the planning and policy branch of the Norfolk Corps of Army Engineers, said she did not think sinking a ship would be allowable under most regulatory rules. Plus, it is very "hard to sink a ship and know what the impacts are going to be." She said there is currently a jetty proposed under "section 107" of a program dedicated for small commercial navigation. Connor said the jetty would help protect the navigation channel, keeping some waves out, but it would not really be a seawall, as some residents understood. A jetty would mostly protect the harbor—the center of the island's economy with its workboats, docks, and crab houses on stilts—from damaging wave energy. Without the protection, the harbor and island could be in peril. It would cost around $2 million. For that kind of money, the island is at the mercy of the federal and state governments, especially the Corps of Engineers.

Connor saw the jetty as a kind of small patch, to begin in 2018, on the way to a much a better plan. For any kind of Corp of Engineers plan, there is the study that could take three years, the authorization that could take another three, and then three or more to build. Connor said they had in mind an even bigger study, of hundreds of millions of dollars in project costs, much more than what Schulte had estimated at $30 million. Whenever they do this, they are required to look at benefits. In Tangier's case, they wanted to look at three main areas: flood risk management, navigation, and ecosystem protection.

Connor said the corps was also looking at some "beneficial use" of dredge material from the ship channel in Baltimore Harbor. It could be used to shore up islands or make small marshes. This has been done successfully for Poplar Island, further up the bay in Maryland. The problem is, the Port of Baltimore has the material and they would have to pay to dump it somewhere else. So Connor said there is a pilot program in the works that may be able to help pay to move that material to Tangier.

The difficulty for Tangier is that other cities up and down the coast compete for these dollars, some with billions in economic benefits, such as the city of Norfolk. But Connor told me that she and many in the Corps of Engineers want to do all they can to help residents of Tangier. "Some of them have given up. And I hate to see that. But I also have to temper their expectations. There's no immediate and quick fix."

Before I caught the afternoon ferry off the island, I went to the museum in the business district, near the souvenir shops. It had a display of land loss. Painted on the wall was a map of the island now, green to show land, blue color of the wall to show water. Sketched in pencil were dotted tracings, the edge of the former island. I asked the two volunteers working there what they thought about sea level rise. The conversation turned to moving. One had heard that FEMA was offering $50,000 to move off Smith's Island, in the same former peninsula cluster just above the Virginia state line, but I could find no record of this. Anyway, it wasn't enough money for either of them to move. How much would be they did not want to say.

Before I left, I ran into the same family that had come into the town hall asking about real estate. We were both getting ice cream at Spanky's, an ice cream parlor. While their young kids licked dripping cones, they talked of a plan for retirement. Maybe owning a bed and breakfast, a sandwich shop. I asked if he was aware of some of the projections for sea level rise. He said he would want to do engineering studies before making a purchase, make sure there was not "beach front property on all sides."

At the counter was a jar for contributions to the seawall, like those at convenience stores asking for contributions to help pay for someone's medical bills. The owner overhead us talking. "Maybe Trump will help us." The prospective buyer/reluctant investor and I looked at one another. I didn't ask on the boat with Pruitt-Moore, but now I had to. "You mean build a wall?"

My skepticism that President Trump would pay attention to the plight of the islanders turned out to be mistaken. In the summer of 2017, CNN ran a story on the island. They interviewed the mayor, James "Ooker" Eskridge, who spoke directly to the president. "Donald Trump, if you see this, whatever you can do, we welcome any help you can give us." He later added that he loved Trump as much as his family. A week or so later, the president called the mayor. "He said we shouldn't worry about rising sea levels," Eskridge told the *Washington Post*. "He said that 'your island has been there for hundreds of years, and I believe your island will be there for hundreds more.'" Eskridge agreed that rising sea levels are not a problem for Tangier—erosion was. "Like the president, I'm not concerned about sea level rise," he said. "I'm on the water daily, and I just don't see it."

Eskridge may not see it on a daily basis, but scientists who study the problem with data sets over a long period of time have seen sea levels rise in the Chesapeake Bay region by as much as a foot (14.5 inches at Sewell's Point), and they are experiencing a rate of rise that is accelerating higher than the global mean. Making a phone call is different from a plan of action.

President Trump offered assurances but no real help, only to cut regulations and the time it takes to study a project. He might have offered to stay in the Paris Climate Accord, which would cut global emissions. Earlier in the year, his budget would have ended federal funding for the Chesapeake Bay program, a federal-state program begun in 1983 and coordinated by the EPA. That money helps protect the health of the estuary where Mayor Eskridge makes a living crabbing.

Astoundingly, 87 percent of voters on the island voted for Trump, a number they are proud of (one grocery store owner has placed that information in large letters on the roof). What is so hard to fathom is that one of communities in the United States whose very existence is most threatened by climate change and its attendant problem of sea level rise supports someone who does not recognize sea level rise or climate change as a problem. The vote tally nearly tracks with that of white evangelical Christians, which is how many on the island identify. And perhaps their concern for how long they have been studied, without action, makes their support for new leadership understandable. The mayor's claim that erosion is causing the problem, not sea level rise, may help him make his case. If sea level rises by several feet, they are underwater. If erosion, the problem is more solvable, even though the forces of erosion would have remained constant over time during the period they have lost two-thirds of their land mass, and sea levels have not.

The owner of Spanky's said they could not afford to be polite any longer. Watermen rarely are. Ferry Captain Crockett would agree. "We have to do something," he told me, looking out on the horizon as Tangier receded from view. He felt like time was running out. "We're not going to let our island go under."

Back in Virginia Beach, Captain Chris Moore motored the *Bay Oyster* over to a dockside oyster bar. It was across from a scrappy clump of sand, mud, and marsh grass, shored up by chunks of concrete from an old bridge, called Fish House Island. It used to be thirteen acres but was one or two now. It's one of the many islands, like Tangier, disappearing because of sea level rise, erosion, and some complicated combination of the two.

At lunch I tried some of the local Lynnhaven oysters, a kind of Atlantic or Virginia oyster (*Crassostrea virginica*). They had a briny taste, a little grit and mineral, firm but slippery smooth with a creaminess. The taste is part texture. Some people push them past the taste buds into the back of the throat, like swallowing a lozenge. I chewed into mine, also put them on a cracker with horseradish, some cocktail sauce, a dash of hot sauce, but also right out of the shell. As connoisseurs will tell you, oysters take on characteristics of where they grow. In the Lynnhaven there is salt, a little soil too, and experts tell me they are fatter higher up in the estuaries than close to the coast because there are more minerals to feed on.

Back at headquarters, while Moore caught up with some other work, Christy Everett, Hampton Roads director for the Chesapeake Bay Foundation, gave me a tour of the Brock Environmental Center, one of the world's greenest buildings. With a combination of solar, wind, and geothermal, a triple threat, they produce 80 percent more energy than they use.

In the foyer, Everett talked about the site, built on what was known as Pleasure House point. I raised an eyebrow at the name, but Everett assured me, in the presence of youngsters, it was not "that kind of pleasure point." The site was a large, undeveloped tract where some developers wanted to build a set of eleven-story apartment towers, but they went bankrupt. The bank had it for $19 million. The Chesapeake Bay Foundation, the Trust for Public Land, and the City of Virginia Beach all partnered and purchased the land for $13 million.

On the wall hangs a giant sculptured map constructed of wood and glass showing the boundary of the Chesapeake Bay watershed. It cuts through West Virginia, Pennsylvania, and all the way up to Cooperstown. Hit a homer and the ball ends up in the bay. Six states and the District of Columbia, where eighteen million live, draining sixty-four thousand square miles of land.

They have no storm water runoff on the site. Rain is collected into cisterns, where it is filtered and treated to become drinking water, the first project in the United States to receive a commercial permit for drinking filtered rainwater. Toilets are composting, waterless units, using pine shavings to cure odor and absorb waste.

To meet the Living Building Challenge (a design framework committed to sustainability), designers did not use any "Red List" materials that include harmful chemicals or materials. Much of the wood was recycled, the flooring coming from a local elementary school that was torn down. The only thing that was not salvaged were the triple-paned windows. The decking was treated with something approved by the Forest Stewardship Council and did not have the usual preservatives.

The facility was built at fourteen feet above sea level, with eight feet of clearance from the ground, ready for a rising sea.

I stepped off the deck to join a few of the forty thousand kids that come through each year. On the day I visited, they were learning to identify land use patterns on the maps spread on picnic tables: marsh, urban, farm/forest, water, beach.

Back on board the *Bay Oyster*, the crew toured some of the Lynnhaven River and a residential area in Virginia Beach with houses built right up to the water. The Neptune festival was coming up, and we cruised past a ten-foot tall painted ceramic statue of Neptune, with crown and trident lit up, his tanned muscular torso emerging as if from the sea, a net-shawl of lights covering his shoulders. He had an outstretched hand, as if to welcome us, or for donations. But a sign warned us to stay clear of the video monitor. Chris noted the development nearby, "as you can see, people continue to build on shoreline."

In the area, new houses were built up on a slope, creating drainage problems for the neighbors. They were not very far above sea level. In Norfolk, they ask for three feet above freeboard for new construction; in Virginia Beach it is two. Thomas Quattlebaum stood in front of the steering console, surveying the scene with folded arms. He mentioned some houses needing as much as $6,000 a year in flood insurance, which "kills the deal." Some areas may have to be let go. No one has identified which ones yet.

Insurance is a complicated piece of the climate change puzzle. We could even say it is at the center of the storm. Most private insurance companies have stepped away from safeguarding against the financial liability of flooding. The stranger, wilder weather and more frequent disasters have made the market too hard to predict. The Federal Emergency Management Agency (FEMA) was designed to be a resource of last resort. They and their National Insurance Flood Program have stepped in where the private companies have left, insuring some $19 billion of home and business value in the Hampton Roads region. Because its premiums have not kept pace with payouts due to disasters, the program is in bad financial straits. They were declared nearly insolvent in 2010, requiring a cash infusion from Congress.

Congress created a law in 2102 to change the program, to make it more actuarially sound, but that would have raised rates. Homeowners pushed back, as did people on fixed incomes. People complained that they were killing property values. So they walked back reforms, back to fiscally unsustainable status quo. Then came Hurricane Sandy, which called for billions more in aid. However, after Katrina and Sandy, the program is $25 billion in the red (before Hurricanes Harvey and Irma).

Even so, the eighty or so private companies that participate in the program did not always honor the full value of the claims. FEMA pays these companies a percentage of premiums plus additional fees to cover their services. They write the policies and manage the claims, acting as a middleman. According to an analysis by NPR and Frontline, insurance companies pocketed some of the premiums they should have paid out. FEMA has found that more than 80 percent of the homeowners who paid into the program were underpaid by their insurance companies. Those companies made profits on average of 30 percent, as much as $400 million in 2013.

FEMA cannot tell people where they can build a home and where they cannot. And people like to live near water. FEMA cannot even tell them how high to build. Only local communities do that. But FEMA provides insurance up to $250,000 (when paid out), which creates some incentive to build on the coast even with sea level rise projections. Some think of flood insurance as a subsidy to build where floods will inevitably occur. According to the Risky Business Project, a nonpartisan effort led by former New York City mayor Michael Bloomberg, former Treasury secretary Henry Paulson, and billionaire financier donor Tom Steyer, $100 billion of coastal property could be underwater, literally, by midcentury.

FEMA provides maps that are intended to show which areas are flood prone and who should buy flood insurance. They updated those maps in 2009, but even those do not consider the effects of climate change or sea level rise, which will dramatically increase flooding. According to the agency's website, more than 20 percent of all claims are filed by people out of designated high-risk zones. If they expanded the map, that would lead to the backlash again. People do not like to be told they live in a floodplain, even if they already do.

In the area we were in, Broad Bay Island, the houses and yachts started getting bigger. The shoreline hardened. I asked Quattlebaum what he thought from a planning point of view. Earlier in the day, Quattlebaum told me he talks to communities about "opportunities," not wanting to emphasize the negative. Surveying the shoreline and houses on both sides, Quattlebaum shook his head: "It's a significant challenge." Said Captain Moore, pointing to another house with landscape timbers at water's edge, falling in, "That is how *not* to develop a shoreline if you want to protect water quality."

They showed me a section of living shoreline, "oyster-tecture" for some. Moore called it a "spat on shell" operation, in which oyster larvae are cultured on shells in tanks before being introduced into the wild. As they grow, they will provide a buffer to wave actions against the shore, "chump change compared with conventional barriers." Meanwhile, they will filter and clean the bay.

Oysters must adapt to very changeable conditions as they are like plants, unable to move. They have developed a wide variety of genes and proteins to respond to changes in temperature and salinity but also different levels of oxygen and pollution. These changes would kill most living creatures. The great thing about oysters as a barrier is that they can adapt to waters disturbed by tides, storms, and other stresses, even to rising sea levels. During the Ice Age, oysters increased the height of their reefs as sea levels rose as much as ten millimeters per year (higher than the current rate of sea level rise in the Hampton Roads region of 4.5 millimeters).

Their hard, calcium-based shells are one adaptation, dissuading octopuses, fish, and crabs. However, the process by which bivalves produce their shells is not fully understood. The Chesapeake Bay Foundation recycles shells from restaurants and from drop-off locations. Once the shells are cleaned, they place them in tanks containing oyster larvae, which once attached to shells are called spat, at which point they go to oyster gardeners or somewhere else in the bay to grow and expand reefs.

Chesapeake Bay Foundation's partner, Lynnhaven River NOW, have installed "oyster castles" in other places. These blocks, made of 30 percent oyster shell, can be stacked like Legos to fit the particular depth and contour of the shoreline. Once oysters attach to them, a reef is born. It was amazing to think that, in the shadow of the largest navy in the world, we are vulnerable to the forces of sea, and that the lowly oyster is among our best choices at coastal defense.

We maneuvered out into a wider section of the river, Wolfsnare Creek, which contains several oyster farms. Farmers lease land for as little as a dollar fifty an acre. The state leases 150,000 acres, 2,500 in the Lynnhaven.

We drove up close to a conveyor belt tumbler on a dock. Oysters could be fed into it and the machine help cracked off the pointy end, making a nice three-inch-sized oyster, what the restaurants prefer to shuck. Each oyster filters fifty gallons of water a day, cleaning out nitrogen, phosphorous, and other pollutants. And since oysters are a keystone species, when they clean the bay the grasses grow, which brings in tiny fish and crabs, and pretty soon things get back to how they once were. Oysters bring life.

A century ago, oyster harvesters in the bay gathered fifteen million bushels a year. In 2005, they were down to 1 percent of that. At one time, harvesters would find a thousand per square meter in the Lynnhaven. Now they find one or two. The bay floor was once covered with them, oysters growing on top of one another like a coral reef. They used machines to dredge reefs so high they would strand ships. Over the past three hundred years, bay oysters have suffered from overharvesting, pollution, and diseases such as Dermo and MSX. But scientists have found ways to grow oysters resistant to disease, and the bay is healthier. Regular rain helps, but too much increases algal blooms. Over the past twenty-five years, surrounding states have spent billions to clean up the bay.

One oysterman was out tending to his cages, holding hundreds of oysters each. His area was marked by poles, like the kind in driveways to guide snowplows. A jet-ski cruised past, kicking up a wake. The oysterman raised a fist. Or maybe a finger. The recreational boater stopped. Words were spoken, ones we could not hear. It seemed like some kind of Western standoff, "I can ride my horse wherever I want." Moore and Quattlebaum told me such conflicts were becoming more common. The harvesting of oysters was not something residents had quite gotten used to, even though it was a normal course of life for generations. Residents wanted to keep the area a private watery playground. For some, those markers and their cages, visible at low tide, marred the view. Looking at the homes on the shore, I sensed a lot of power and money that could work to protect that view.

We were now off the coast of First Landing State Park, named for the English colonists who arrived in 1607. When John Smith roamed the shores, he famously found oysters as big as dinner plates and said they could feed a family of four. Though his writing was often full of false swagger and self-promotion, he was apparently not bragging. Locals have found shells as big as hub caps.

Those settlers arrived on a hot, swampy peninsula. They showed little ability to gather or grow food, to hunt or fish. They did not adapt well to the climate. They fought among each other and failed to make proper seasonal provisions, leaving them hungry. Early in their settlement, natives shared food supplies and knowledge of how to plant squash and beans, how to make weirs to trap fish in the creeks, how to hunt deer, turkey, and other game.

According to Francis Jennings in *The Invasion of America* (1975), the official instructions of the Virginia Company referred to natives as "natives," "natural people." In later publications, they became "savages" when the necessity for defense was mentioned. "John Smith, for all his swagger and bluster, was ready to regard the Indians as peers when the price was right." As historian Paula Marks Mitchell writes in *In a Barren Land* (2002), when the British came as traders, they valued natives as the French did, but when they wanted to settle and colonize, things became different. They wanted what the others had, especially land. Skirmishes increased as hunger rose. And unlike the natives, the Europeans oriented around one permanent location, while the natives moved from place to place depending on the season, though they too, adapted to the European custom because they had to: others were taking their traditional grounds. Narratives were developed to justify behavior, about one's superiority over another. Providence or the name of God was invoked to excuse it.

In our national myth, these brave, hardy, self-reliant settlers survived because they got down to work, ordered strict discipline, pulled themselves out alone. But really they were radically interdependent, on natives and each other, as were crops of the soil, the oysters of the ocean.

In *Generall Historie of Virginia* (1624), Smith himself promoted another myth: that settlement would be somewhat easy. It would require "industry," but conditions were right: "The mildness of the air, the fertility of the soil, and the situation of the rivers are so propitious to the nature and use of man, as no place is more convenient for pleasure, profit, and man's sustenance, under that latitude of climate."

The settlers found the climate too hot in summer, too cold in winter. Tree ring analysis shows they were in the middle of an extended drought, and evidence from archeological digs shows that 1607 fell within a cool period that climatologists and historians call the "Little Ice Age."

That climate is changing, in effect moving the latitude. Under the current rate of emissions, and warming, experts at the Nature Conservancy have created maps that predict Virginia will have a climate more like South Carolina by midcentury, and one like Florida by 2080. Average annual temperatures will continue to rise, making the climate of the new "Virginia" something like that of the tropical zones.

To some, that might sound appealing. Palm trees in Richmond. Short sleeves all year-round. But such changes will cause huge disruptions to ecosystems and food supplies. According to the Centers for Disease Control (CDC), more people in the United States die from extreme heat exposure than from hurricanes, lightening, tornadoes, floods, and earthquakes combined. Our cities will be hotter, causing changes we will have to adapt to. Alternatively, a wise approach would be to forestall the heat as much as possible by curtailing heat-trapping gases in the first place. That would require international cooperation.

When studying oysters, a German zoologist developed ideas about the cooperation of living systems. In the 1870s, Karl August Möbius was asked by the agricultural ministry in Prussia to study artificially cultivating the oysters beloved by elites. He concluded that they could not: that the right conditions of temperature, salinity, and mineral content were too difficult to replicate. But while working with banks of oysters, he developed a notion important to the emerging study of ecology, the living community, which he called *biocoenisis*. It comes from *bios*, life, and the Greek *koinoein*, to have something in common or share. Every oyster bed, he wrote, is a "community of living beings, a collection of species and a massing of individuals, which find here everything necessary for their growth and continuance." Once the oysters were reduced, the balance of the community shifted to something else, like mussels, which took the oyster food. He applied the theory to other things, such as the extinction of the dodo. Pigs and other animals "disturbed the biocoenisis of the island."

The truth is, no place is an island. We're on a planet of connected systems. What happens in one living community affects another.

Tangier is affected by geologic processes begun millions of years ago and happening thousands of miles away. Thomas Cronin, a senior geologist at the US Geological Survey who has studied climate change and sea level rise in the Chesapeake Bay, told me "the contribution of the melting ice is unquestionable." For him, the meteor theory is "bogus," although the melting of the Antarctic ice sheets, as old as the meteor, are a grave concern. As to why the region is experiencing slightly more rise due to subsidence, he finds good evidence for the rebound effect of past glaciers that once pushed the coast up. On the other side of the ocean, he told me, northern Europe is experiencing the opposite but related effect, an uplift.

After our tour of the Lynnhaven, Captain Moore guided us back to the dock at the Brock Environmental Center. The kids had been released from the picnic table lesson and were scavenging the shore for shells. We had been off the shore of First Colony State Park, the most visited state park in Virginia. Nearby is also the least visited, False Cape State Park, accessible only by boat, bike, or foot. It was called False Cape because sailors mistook it for Cape Henry, the mouth of the bay. Rather than a safe harbor from the turbulent Atlantic, they were lured into the shallow waters where they often ran aground.

Drifting into the dock, I glanced up at the center, built on a point. I thought of it as a kind of lighthouse or beacon, guiding the way for the sixty-four thousand square miles of land that drained into this bay. As a center for education, the Brock aimed to be a guiding light. From the moment the kids step off the bus and step onto the permeable pavers in the parking lot, the kids receive a whole systems education: about the connections between storm water and the health of the bay, energy use and its effect on a warming planet and rising seas, the way a watershed connects us, unites us, how what happens upstream affects those down.

I poked around with the kids along the shore for a little while, thinking of what it would take to restore the bay to its former state, to how John Smith found it, how the Powhatans knew it. With the right policies, future generations would be able to enjoy it. Then I said goodbye to the *Bay Oyster* crew.

As I walked back to my car, I caught up with a ten-year-old boy, Jackson, whom I had seen at the day's lesson. When asked to find "the important parts" on the land-use map, in terms of filtering the bay, he got the answer right and pointed to the marsh. Other kids were trying to find their house. Developmentally, it can be a hard lesson to learn: that the world does not radiate out from *me*.

By the driftwood gateway arch that welcomes visitors, I asked him if he enjoyed his day. He told me he did. I remembered how I loved field trips when I was a kid. Especially the one to Sandy Hook, New Jersey, to learn about horseshoe crabs, those blue bloods. He asked me where I was going.

"Home," I said. Though I felt the need to offer that I lived in a different watershed.

"I live in this one he told me," making clear he had absorbed the day's lesson. "This is my home." He was referring to the ground he stood on, at the mouth of the bay, the ocean beyond it, under driftwood from near and far, a piece in a much bigger ocean universe. An oyster.

3

The Proximity of Far Away

Climate Change Comes to the Alligator, North Carolina

March 1. By now my family has usually heard a few spring peepers, high, tinny bell ringers, in the marsh across the creek from where we live in Virginia's New River Valley. A few individuals emerge first, and by mid-March their singing rises, at dusk, to a full crescendo. Winter has had a firm grip on the East Coast this year, so they are at least a week late.

Like a well-known Concord naturalist, I have been keeping a record of the first spring peeper in our marsh. A meticulous notetaker, Thoreau recorded the date ice disappeared from the middle of Walden Pond, the ice-out date, in his journal from 1846 to 1860. The average date for the fifteen-year period was April 1. More recently, volunteer students from Journey North and park rangers at the pond have made note of the same event. Since 1995 (throwing out 2010, when the ice never came in, though it was sixteen inches thick in Thoreau's time), the date is around March 17, a full two weeks earlier. We know what's going on. We know why. But why is this information not connecting to people and their actions?

Those peepers are responding to their circumstances and environment, heeding an ancient call for survival. The ground is mostly unfrozen (though they can survive if it freezes again), the days longer, and the males try to outdo one another in song. Called "pinkletinks" on Martha's Vineyard, "tinkletoes" in Canada, they are small enough to fit on a fingertip, but man, can those frogs make some noise.

When those peepers begin to call, something is changing in me as well. I am (almost) ready to put away my skis and usually feel an urge to plant something—tubers, greens—in the soil. Warmer days and the return of light, the sap starts moving in us, spring fever. The oldest part of our brains, which evolved some two hundred million years ago, is often called the "reptilian brain," and we share it with frogs and birds. It controls life functions such as breathing and heart rate but also our fight-or-flight mechanism. Lacking language, its impulses are instinctual, ritualistic, concerned with survival. The basic ruling emotions of love, hate, fear, lust, and contentment coil up through this first stage of the brain. Over millions of years of evolution, layers of more sophisticated reasoning have been added to it, can override it, but it gets first call.

The climate crisis is global, but the effects are felt most intensely as place based and local. Between the disasters—the droughts, the hurricanes—are the early-blooming orchids, the thin layers of ice, the early arrivals of spring peepers, the late departure of migratory birds. Noticing these small changes requires knowing a place deeply, the kind of local knowledge that is passed down over time. Surely the ability to recognize patterns is also evolved into us. The hunter sees a deer on a particular path and might expect that deer again. Such skills in pattern recognition also keep us safe from lurking dangers.

The ability to mark small changes unfolding over time is lost to many of us living in a harried, frenzied now, with a short attention span news cycle. Climate change is, however, the inevitable effect of past actions not only on the present but on the future. The patterns and trend lines are there, but a series of severe storms does not a climate change make. Unless linked to particular disasters—and scientists warn against it—climate change does not make great TV news nor does it seem to rouse many people from their torpor to move on the issue of global warming. No "sufficient catalyst" has yet occurred to change the tide on climate change, though the ocean tides are rising: not the BP oil spill, the warmest decade in history, Superstorm Sandy, or the droughts in California. If climate change were a war, there has yet to be a Pearl Harbor.

At one time, environmentalism could afford to be intensely local. It was Thoreau fishing his pond, Edward Abbey picking Kleenex out of cliff rose and cactus, Rachel Carson documenting the effects of DDT on local communities, right down to the very worms. As Adam Rome shows in *The Genius of Earth Day* (2013), the early environmental movement also consisted of suburban mothers who wanted to protect their children from the immediate threats of contaminated air, water, and food. The danger of global warming, caused by an invisible, naturally occurring gas, has been less local and seemingly less direct in its impact.

More and better data could win over some climate change skeptics. We have maps of the arctic ice sheet that show how the ice is contracting on the surface, but geophysicists also show how it is melting on the bottom, decreasing in thickness. And while most discussions of global warming focus on the air temperature, about 90 percent of the heat generated by the greenhouse effect is warming the oceans, according to the Intergovernmental Panel on Climate Change. Oceans are also acidifying as a result of increased carbon dioxide, impacting wildlife and habitat, phenomena that can be measured but hard to capture in a picture or map. I imagine some webcam or drone that could give a "real time" perspective on the changing climate, like the popular wildlife cameras, but the overall effects of climate change are too slow moving, especially as they impact our lives.

We have good pictures and data points. We see ice caps calving and glaciers retreating. We have maps that show projections of storm swell and the low-lying areas that are drowned by sea level rise. What we don't have is global action.

A 2015 poll by the Yale University Program on Climate Change Communication found that, while 70 percent of Americans say that climate change will affect future generations, only 41 percent think it will affect them personally. Though a good deal of the skepticism could be attributed to the doubt sowed by energy companies and their lobbyists, perhaps something more is going on, something the ideal of a global imagination may fail to account for.

In *Don't Even Think About It* (2014), George Marshall discusses some of the science behind climate change denial. He surveys the work of psychologists, such as Daniel Gilbert, who argue that we evolved to respond to more immediate, abrupt, and direct threats than the gradual, distant one of climate change, at least in terms of its eventual magnitude and effects on our personal lives. Addressing it requires making sacrifices now in order to prevent unclear costs in the far-off future, which humans (and their elected officials) are not good at. And there is no clear enemy, except you and me and everyone we know. Although Adm. Samuel J. Locklear III, the commander of the U.S. Pacific Command, has said that global climate change is more dangerous than terrorism, our reptilian brains respond more to the terrorist threats.

Trying to make sense of the lack of urgency, I wondered if we should see more response, less denial, among people in places most affected by climate change. After all, for certain low-lying areas climate change is not just theoretical but a direct assault on their property and place.

I sought out people on the Outer Banks of North Carolina. On the weekend I went—the last in August—the National Weather Service map of the United States showed little activity anywhere. It uses colors to show weather events, but the country was unshaded, showing only the colorless grids of county and state lines, except for a magenta section of eastern Washington, a red flag warning for wildfires, and most of Idaho was shaded in gray, an air quality alert. Then in that small bulge of North Carolina that juts out into the sea, a pale green flood advisory. What better weekend to go? I ignored the advisory and packed my gear. After all, it was just a representation of a flood, not actual water lapping at my tent.

My day started with a walk to the beach, but at this beach access, near the Black Pelican in Kitty Hawk, there was little beach, only a small area between waves and a massive, man-made dune. The dune was a form of "beach nourishment," where sand is pumped or dumped, not really stopping erosion but giving the waves something to chew on for a while. A few gulls wandered along the wrack line.

Then to Ten O Six, a "beach road bistro." The owner, Toby Gonzalez, wearing a white, stained T-shirt, emerged from the kitchen with a good cup of coffee and an even better breakfast burrito. He had lived here his whole life, so I asked him about changes to the island, especially related to sea level rise. He said he hadn't really noticed any but that it was "hard to argue with the numbers." Still, the numbers did not seem to convince him. I think he said this for my benefit, perhaps a nod to my being a patron.

Another customer came in, another lifelong resident, from the northernmost town, Duck, and Toby asked him about changes. "Tide comes in," he said. "And tide goes out." They laughed. I asked about projections of sea level rise up to three feet by the end of the century. "I won't be here," new customer Dave said. And Toby started riffing. "If it does rise, I'll have to charge more for my sandwiches," he said with a wink. "Beach front property." "Damn Atlantis," he went on, smiling, "trying to claim our women."

According to a report from the Yale Program on Climate Change Communication begun in 2008 and updated yearly, there are "six Americas" when it comes to perceptions of the impact of global warming: alarmed, concerned, cautious, disengaged, doubtful, and dismissive. The comment about the tides coming in and going out seemed to come from the perspective of the doubtful, and the comment about it being "hard to argue with the numbers" seemed cautious. And while the jokes would seem dismissive, I detected a bit of "concern" in them. Global warming is happening but is still a distant threat. While he lingered, Dave from Duck added, "We keep fucking with this planet, Mother Nature is going to do something." But his main solution was more sand fence. He pointed to a picture hanging on the wall of a fence halfway buried by a dune.

I told them I was headed to the Alligator National Wildlife Refuge area because they are already seeing evidence of sea level rise. Dave from Duck said some of the land there he used to hunt on "floats." I couldn't tell if we were having a problem communicating. Floating land? Was Dave floating on something when he hunted? But then I remembered that much of the land in the "Alligator" is peat, spongy. Christine Pickens, of the Nature Conservancy, later told me that indeed, in parts of the Alligator there are sections of marsh and vegetation that float, not unlike mangrove islands. They are tricky to walk on. Step through and beware of what lies in wait below—gators.

Before heading east across the bridges to the mainland, I stopped at Jeanette's Pier, where concrete has replaced the wood wiped out by Hurricane Isabel (2003). Owned by the North Carolina Aquarium Society, the pier focuses on education about fish, marine science, and resource conservation. The parking lot is made of porous pavers to help with flooding. On that day, they were holding a surfing tournament. A few surfers were putting on wetsuits and readying their boards. I asked if sea level rise would affect their sport. Both were "doubtful" it would, but neither did the loud music and promotional banners lend toward reflection.

Part of the reason I wanted to visit the refuge is that according to a 2005 report, it has a higher density of black bears than any place on the East Coast, three per square mile. Also, it has a population of red wolves, which I didn't even know existed in the East. I met up with the refuge manager, Mike Bryant, and the wildlife biologist, Dennis Stewart. Mike is a fit fifty and wears gray stubble. Dennis is taller, closer to retirement, and has been with the refuge since it was established thirty years ago.

In a conference room at the center, I pulled out my map of "sea level rise vulnerability" in the Albermarle-Pimlico Estuarine Systems, trying to get oriented, trying to figure out what's going on. Most of the land is red, "severe" vulnerability, according to our key (less than half a meter), with a few moderate splotches (over two meters).

My map represented a "bathtub model," visualizing water at a constant elevation. Right away, Stewart pointed out how the yellow hummock is misleading because it is a peat mound, a mound of decayed vegetation. If sea level started to rise, it would degrade the peat. Peat accretes faster than it decays under normal conditions, but if it becomes saturated with salt water or overly dry and exposed to air, it breaks down and looks like "coffee grounds."

Stewart talked about how the refuge was once a healthy forest of cypress, black gum, and pond pine, or pocosin pine, found all along coastal marshes. But lately, tree mortality has risen dramatically due to increased saturation and salinity. Stewart saw a "wedge of change," where salinity creeps in and attacks one specific place and fans out from there, attacking species that are not salt tolerant, like the trees. The effects of salinity are particularly bad near some of the ditches, or sloughs. These were often created to make roads, with material being dug out of the system to elevate the road. Now these sloughs carry some of the salinity from the rising sea.

The forest is a kind of frontline community. But due to the rising sea, the pines are dying, transitioning to shrub, and they are giving way to marsh. "Where there was forest, there's now marsh." Bryant used an analogy of placing coins under a table leg. We may not notice the change; it is subtle, but it is there, and he is hoping to forestall it. "We're not gonna stop sea level rise but want to create some resilience," especially for the creatures who depend on the refuge, like migrating waterfowl. Christine Pickens told me something similar: "We are not going to stop the change, but we might preserve function for as long as possible."

At the refuge, in concert with the Nature Conservancy, they create oyster reefs off the shore to slow the rate of erosion. The reefs slow the wave energy before it reaches the shoreline. Pickens told me they measure erosion on the shore both where there is a buffer and where there is none. Erosion is 25 percent less with oyster reefs, and 50 percent slower when they use marl, a lime-rich mud, embedded with oyster shells. Pickens and the Nature Conservancy have also put check valves in some of the sloughs to reduce the amount of salinity brought in during floods.

Some of the traditional fixes to shoreline erosion are to install bulkheads, walls made of rocks or wood to stop erosion. But barrier islands have complicated dynamics. They migrate and move. In a process known as barrier island rollover, some of the sand on the coast rolls over the other side and eventually feeds the marsh. The wind comes in from the west, sometimes bringing sand back. By building up the shore, we actually hurt the natural process, and North Carolina loses about three feet of shoreline per year. Engineering projects meant to protect human activities, such as dredging and building hard structures called jetties and groins, have made erosion worse. Without human interference, the islands would adapt to accelerating sea level rise by migrating west, says Duke University's Orrin Pilkey, an emeritus professor of Earth sciences. Instead, he told *National Geographic*, they are "standing perfectly still, and we're beating our head against the wall trying to hold those shorelines in place."

The response on the part of some of the beach communities is roads and engineering solutions first, and when Dennis and Mike have advocated for the wildlife or for solutions that work with nature, they are met with resistance. The piping plover? "How does it taste?" said one group. In the politics of the seashore there are the "from heres" and the "come heres," and they often distrust one another.

The North Carolina Coastal Resources Commission, which regulates land use in the state's twenty coastal counties, asked a science panel to assess their vulnerability to sea level rise. The panel reviewed the scientific literature and projected thirty-nine inches of rise before the end of the century. Roads would be underwater. Properties would have to be abandoned or moved. Money, a lot of money, would be needed. A member of the science panel, coastal geologist Stanley Riggs, said the islands will soon resemble a "string of pearls" separated by shoals unable to support a fixed highway. The twenty coastal counties and many realty companies complained. The report scared people, especially tourists. So the governor appointed a new commissioner, Frank Gorham, an oil and gas man who announced that the panel must limit their study to a thirty-year projection. The model for sea level rise did not change, just the timeline. A one-hundred-year forecast, said Gorham, lacks credibility. We agreed it was a Scarlett O'Hara moment: "Oh, I can't think about this now! I'll go crazy if I do! I'll think about it tomorrow."

"The main problem they have is fear," Michael Orbach, a marine policy professor at Duke University, told the *Washington Post*—especially fear of damage to the coastal economy. Something threatens what they hold dear, and the amygdala activates. But it's an old part of the fear center, one motivated by a different kind of survival, often an economic one. Ideology drives this fear —"motivated reasoning," the social psychologists call it—and facts and evidence have a hard time trumping it. These people lie mostly on the far end of the "six Americas" spectrum, dismissive. Use the words *climate change* with such groups and prepare for resistance. The reptilian brain throws up defenses. "It's about how you frame it," one wetlands conservationist told me. When he first talked to local community groups, he talked about climate change and showed maps of sea level rise, but those efforts went nowhere. Pictures were more effective. He would show a stump sticking out of the water, evidence of saltwater flooding. Someone in the audience would recognize the place: "I used to hunt there."

What is very difficult to communicate to such groups is how their self-interest should involve long-term planning. The piping plover is a surrogate species, one of the many in the history of environmentalism—spotted owl, polar bear—that would suggest to some an interest in valuing species over human communities. But they are often a dying canary in a very dangerous coal mine. Working to save the Alligator Refuge is working to save a possible source of fish, game, clean air, and clean water. The alligator in that refuge are at the northernmost end of their habitat. A change in temperature, or salinity, could drive them out. They control the population of small mammals, which feed on wetland grasses, which control erosion. Without alligator, less and less Alligator.

Trying to learn more about tree health of the state farther inland, to the mountainous west, I met with Chris Oishi, research ecologist with the Coweeta Hydrologic Laboratory at the U.S. Forest Service in Otto, North Carolina, near the Georgia border. Chris earned his PhD at Duke studying carbon and water cycling. He explained to me that the early researchers at the laboratory tried to understand how foresting and management practices affected stream health. More recently, they have looked into how climate affects hydrologic processes and the forest.

In the 1930s, researchers studied chestnut blight. Today, it's the hemlock wooly adelgid, which was brought into the United States from Asia accidentally and discovered near Richmond, Virginia. It has now been seen nearly the length of the Appalachian Trail, from Georgia to Maine, causing widespread mortality of hemlocks. Adelgid do not overwinter well, so their range stops at a certain point, but the length of time it stays warm in the south extends the growth cycle, where they can do more damage. When the chestnuts died, species filled in. But nothing, as of yet, is replacing the hemlocks, which grow near streams and cool valleys, a specialized habitat. The loss of these hemlocks has implications for carbon storage and fish health. In the short term, Chris explained, "the fish love the deadfall," as they create pools and dams, add nutrients, but eventually, the overall quality of the stream will suffer.

Since the 1980s, the laboratory has seen annual mean temperature increase in both the dormant and growing season. They see an increase of 0.5°C (0.9°F) per decade. On the day I was there, it was in the midnineties. Chris told me to bring a rain coat, but that day was warm and dry and had been for a few weeks.

At an observation tower, an automatic camera takes pictures, and when Chris and other ecologists compare these to other shots taken years ago, they see "bud burst" about two weeks earlier, increasing respiration in the trees, which releases carbon. One of the other trends they have noted is climate variability. The wet years are wetter, and the dry dryer.

Researchers thought that with heat and drought, the oaks would fare better, as they use less water than do maples and poplars, but "oaks aren't doing as well as predicted." One hypothesis on why the oaks, which are fire adapted, are not doing well has to do with fire suppression. At one time, southeastern forests caught fire from lightning strikes (or from Native Americans).

Chris, tall and slender, as at home on a trail run as modeling carbon cycles, took me out to see one of the research stations. We donned safety vests, in case of hunters, and hard hats, for deadfall. Especially from hemlocks.

In something that looked like a bee hive, they recorded the maximum and minimum temperature on strip charts, also reading humidity, since both affect the water taken up by trees. Precipitation gauges showed when and how much rain during a period of time. Scientists could look at the inputs (rain) and outputs (streamflow), which would tell them what the forest uses and anything that evaporates.

Chris described this hydrologic system as a "black box," where they could measure effects and variables. In one early experiment, they cut the trees in the watershed, planted white pine, and measured how much water the forests used. The pine plantation used about 40 percent more, since they are fast growing and evergreen and have more overall surface area.

After crossing a stream, we arrived at another station, the "electric forest," a wired system used to take the pulse of the trees and hydrologic system. There were two stations: a tower, like a cell phone tower, and one lower to the ground. Both were placed in a deciduous forest common to the Southeast: white oak, tulip poplar, red maple, black gum, and patches of rhododendron.

Over the sound of cicadas, at the base of tall oaks, Chris explained some of the work going on there. Laundry baskets were put out as "high tech leaf litter collection." Something like cheesecloth over the top collected what came from above. Leaf litter tells of "pools of carbon." Under some foil were sap probes, used to estimate flow rate of water moving through tree. Not unlike a doctor measuring blood pressure, from these probes they can determine the health of the tree, how much photosynthesis occurs and how much carbon is taken in.

At what looked like a TV antenna, Chris showed off an "eddy flux system," a device for measuring the "net exchange of atmospheric gas across a horizontal plane." Basically, there was a vortex of gases, like in an eddy of water in a river, or milk in coffee. In this case, it was carbon dioxide being burned off from the floor and from above. Ten times every second, the instruments measure wind speed and direction, updraft and down, different concentrations of gases, water vapor and carbon dioxide entering or exiting the system. "We get a sense of the forest breathing. At nighttime we see carbon being sourced, during the day a carbon sink."

According to researchers at the laboratory, in a highly productive year, the amount of carbon dioxide removed from the atmosphere and stored in an acre of forest is "roughly equivalent to the annual emissions of one automobile."

The warmer temperatures as a result of climate change can increase forest growth but also increase the loss of carbon captured because of increased respiration. Trees both absorb carbon during photosynthesis and give off some during respiration. As Chris explained it, respiration (the cellular-level process of breaking down sugars to release energy that consumes oxygen and releases carbon dioxide) from living cells increases exponentially with warmer temperatures. Therefore, on a warm day, tree leaves, branches, trunks, and roots all release more carbon dioxide than on a cooler day. Similarly, microbes in the soil will break down (decompose) organic matter in soil at a faster rate during hot days. However, Chris pointed out that warm temperatures could also limit photosynthesis, if conditions are too hot or dry. If the trees die, like the hemlocks, less and less carbon will be absorbed, meaning planetary warming will increase.

At Coweeta, they take multiple small-scale measurements and look at the conditions that made them happen, so they may be able to make predictions about where we are headed.

After the tour of the "electric forest," I drove up on roads built by the Civilian Conservation Corps to find a place to hike, to breathe amid the trees breathing. On top of the ridge, the temperature dropped fifteen degrees. On the way up, I noted more than a few dead hemlocks. Walking along the trail, the air felt conditioned, leaves blocking sun, water vapor absorbing heat energy.

I went to Pickens Nose trailhead because I heard there were falcons nesting in the cliffs. There might have been, but I stumbled into a group of adults learning to rock climb. They had come from Western Carolina University, and the instructor was trying to teach an older man to rappel.

To do so, he had to lean out, perpendicular to the rock, using the rope to lower himself and keep his feet on the face. But that requires a leap of faith, of trusting the rope over your instincts. The instructor kept saying "lean out, sit back," the rappeller kept going down but not out. His feet wanted to, but his brain would not let go.

Before I left the Alligator, I wanted to paddle in the refuge. I put my kayak in at Milltail Creek, slipped out into a bigger lake, and then floated into a narrower stream. A slight breeze came up, and I enjoyed the sound of the water lapping against the boat, the wind whisper through the grasses and trees. I heard the twangs of a few bullfrogs and the calls of birds. I spied an osprey, big and graceful, and scanned the shore for bears or wolves. I thought of howling but didn't want to disturb the calm.

I stayed in the middle of the creek, avoiding the floating grasses and the few logs sticking out from shore, although there was really no clear border, no firm shoreline. In a way, no map. I paddled on, enjoying the scenery and then a boil of tannin-colored water and tail. I saw it. It was only six feet away and maybe that long. I saw it over my exposed legs and feet, for I was in a sit-on-top kayak. The alligator had a protective armor, tough skin and spiky plates. I had a tippy piece of plastic. It was there in a flash and then under water, but I caught a glimpse in the moment my heart stopped, of a nightmarish claw, serpentine tail, the jawline. I may or may not have seen teeth, but teeth are beside the point. They came with the body.

Kayakers are supposed to paddle through turbulence, keep the momentum going, but while my heart raced, my arms no longer moved, stayed frozen as the boat drifted past the alligator place, as I took in what I just witnessed.

The earliest of European explorers to encounter the alligator in the New World were so awed by what they saw it summoned hyperbole and inaccurate representations. In *New Voyage to Carolina* (1709), John Lawson says the alligator "sometimes exceeds seventeen Foot long. It is impossible to kill them with a Gun, unless you chance to hit them about the Eyes." He claims that "against bad Weather" "they roar, and make a hideous Noise." William Bartram notes their roar in his "Battle Lagoon" passage of *Travels* (1791). The alligators threw his "senses into such a tumult." After watching them feed on fish, "shocking and tremendous," he found he could barely sleep and was unable to suppress his "fears and apprehensions" of being attacked. One alligator rushed out of the reeds, and "with a tremendous roar came up," darting under his boat "as swift as an arrow," then "belching water and smoke that fell upon me like rain in a hurricane." In both passages the alligator is associated with weather: both, it seems, were difficult to predict or know.

My alligator went underwater without a roar, but my kayak had to pass over the spot where it had recently lurked. I didn't expect to see an alligator, but now it was eyes everywhere. I had planned to paddle all the way out to the Intracoastal Waterway, a few hours, but I turned around after another thirty minutes. I was hypervigilant, but I didn't see it again, until I did. On the way back, a set of eyes on that bridge of the forehead emerged just above the surface, spied me, and then disappeared. Again, I had to cross its water path to get where I was going. He had certainly increased the activity in my adrenal gland and in my reptilian brain. I was ready to fight if it came to that. I had a firm grip on my paddle, a possible weapon. I also told myself all the things the frontal cortex tells you, the thing that says "it's only a movie" or "it's more afraid of you than you are of it."

At the time, I was going over the day's notes in my head and looking for an ending, but "leg gruesomely ripped off by massive reptile" was not it. Neither was "parts of professor's body found floating in lake." But what a rush. Like climate change, the gator is a thing that can't quite be controlled. And when it vanished, I thought of how in matters of climate change, and reptiles, it's that thing you can't see that can kill you. I was ready to engage it if I had to, and if people in the doubtful-dismissive zone felt the direct presence of a climate change threat, they would too. For all our models and projections, to really know what is happening, it may be best to see changes (and make measurements) in the field.

My wife and teenage kids teased me when I got home. "I saw an alligator," I tell them, a fish story hard to believe in the Virginia mountains where we live. "What did you expect?" they say. "You were visiting a refuge called Alligator." Only I didn't expect it. I read that it was a possibility, but some representations lie to us. Some maps do too, or we lack some kind of spatial and temporal understanding to grasp how the models on them really impact us. All the climate change maps and models predict some pretty bad things, so bad that an understandable response is resistance. Back people into a corner, and they want to fight back in some way. The alligator gave me a way out. And then it dove deep down below into the darkness.

Darkness is one way this story could go. Nearby the Alligator is Fort Raleigh on Roanoke Island. It tells the story of the Lost Colony. In May 1587, in the time of Shakespeare, Sir Walter Raleigh sent John White with three ships carrying eighty-nine men, seventeen women, and eleven children to set up a permanent English settlement. Although intending to settle near Chesapeake Bay, the settlers were forced to locate on Roanoke Island in late summer, with little time for planting. Virginia Dare, named for the Virgin Queen, was the first English child born in the New World, but the community had a difficult time. White returned to England for more supplies, but the Spanish Armada tied up shipping. He did not return until August 1590 and found no trace of the colony, only the word *Croatoan*, the name of a local Native American group, carved on a tree. This seemed to indicate the colony had moved to a nearby island called Croatan (now Hatteras), but storms prevented investigation, and White returned to England without ever knowing what had become of his family or the colony. One explanation for what doomed it: climate change. The archive of tree ring analysis on bald cypresses shows that the worst drought in eight hundred years occurred between 1587 and 1589.

Of course, climate always changes, but the rate of change has accelerated now. Without the resilience to adapt to the changes, like that of the Native American communities at the time, the colony was lost. Our colony could be again now. But there is another way the story can end. Several miles to the east is another National Park Service exhibit, that of the Wright Brothers, who relied on teamwork and an application of the scientific process, along with close observation of what was happening in nature and bird wings, to fly.

Much of the climate change news is dark, so dark it can lead to despair. In dim rooms at conferences I have watched presentation after scientific presentation showing graphical projections of carbon dioxide and temperature and their likely effects. Yet, too often the graphs and data and manner in which such presentations proceed fails to speak to something deep within us, something primitive: a need to survive. The alligator can live in that darkness, but we cannot. Despair for the dying patient does not lead to rescue.

Navy SEALs are said to undergo a test in which they use SCUBA gear underwater for twenty minutes. Then someone ties a knot in their tubing, activating the fear response. The reptile brain shrieks. The candidate can't breathe. Most give in to their fears and swim toward the surface, but some set goals: untie the knot. The amygdala screams, "Breathe! Get away from the thing trying to kill you!" But others quiet that part of the brain, and they execute. They plan, and then, they come up toward the light.

In the Alligator Refuge, they note the small changes on the ground, the changes that signal bigger ones to come. They note those changes and design strategies, with nature in mind, to reduce the impact of climate change. For most of us, there is no direct tangible experience of an ocean that is more acidic or a sea level rise that has yet to occur. For climate change to resonate with people, it must be proximate and present, not only general and global but also micro and immediate. To save the Alligator, we will have to engage that part of us we share with the alligator, that protective instinct, and also that very human part of us, language, that allows us to tell stories. Those stories should involve how climate change affects something very near and dear to each of us until they make a chorus of voices, no longer dismissed or ignored. Peepers, relatives of alligators, come out of the dark ground in spring, singing, "The Earth is warming up." It's time to get a move on.

4

Fish out of Water

High Tide in the Lowcountry, South Carolina

After the storm, people returned to the city and assessed the damage outside their homes, shielding their eyes from the morning sun. They picked up sticks and chairs blown over. I was driving around the battery area of Charleston, a promenade named for the Civil War coastal defense at the site. It stretches along the lower shores of the peninsula where the Ashley and Cooper Rivers form Charleston harbor and is one of the more picturesque scenes in America. Across the harbor is Sullivan's Island, a land mass with a lighthouse and a water tower. On the morning after Hurricane Matthew, the sun was out, the wind and rain gone, and I talked to residents cleaning up. They were sweeping the leaves of palm trees and other debris into piles, and I was checking out the puddles left over from the rain and the storm surge.

Then something moved. A fish moved. A fish that had been brought over the seawall and was now on the wrong side of it, slithering along the edge of a curb.

I chased after it, unsuccessfully, and a man on a bicycle stopped to take a video. "This is a good deed, man," he said. "This is who we are as human beings." I kept after the mullet, swimming in and out of my feet, impossible to grab. I hoped to beach it on shore and then trap it so I began splashing. "This is the new Charleston," my new friend said. "We've come a long way from a slaver's town. We're the type of people who will stop to save a fish." I finally grabbed it, held its wriggling tube of muscle with two hands, walked it up over the wall, morning joggers now pausing to look, and released it back into the water. After the storm, much else would need to get back to normal.

My narrator wore cut-off shorts, a bowler hat, blue John Lennon sunglasses, a long beard, and an old suit vest over a red T-shirt. "All right, man. I know your hands are nasty. Let me shake your hand." He is Larry of Liberty Carriage tours. And he was talking to the assembled group now, "Now if you're a Buddhist and you believe in reincarnation . . ."

He gave me a little of the tour, telling me about the historic buildings behind me, one believed to be the first hospital in the Carolinas. Another was owned by a man who the character Rhett Butler was based on. That played well to tourists, I was thinking, but Larry seemed knowledgeable. I later looked it up and it was true: Rhett Butler was from Charleston and his character may have been based on the real life of George A. Trenholm, a wealthy blockade runner who lived in Charleston. From the piazza of the house behind us, Gen. P. T. Beauregard watched the bombardment of Fort Sumter, signaling the start of the Civil War. The houses were old, well-preserved, Greek Revival with tall pillars and high wrought-iron fences.

Larry free-associated on the fences. Tall ones were built for protection. Before the Civil War (and until about 1920), the majority of Charlestonians were black. A tall fence kept potential "marauding slaves" out. The low, wooden fences meant the iron was likely melted to make bayonets and cannon balls during the war. A low iron fence, built in its place, was a postbellum fence, more decorative, meaning residents no longer lived in as much fear.

Larry felt like it was a new day in Charleston. They still looked to the past, but they were forward thinking. I was there to see how they were faring from the storm. I told him about the project I was working on about climate change. Larry said there are two kinds of people, those who "run at a challenge, and those who run away from it." All I had done was save a fish, I said. "But you did something while others looked on." He felt like that was what the city was doing now, acting and not watching.

Larry said the city used to experience bad flooding in the spring, water barreling down rivers, inundating the farms and forests. Eventually, and often through slave labor, they drained and diked flood-prone areas, but much of the area we were in was built on marshy fill, even lower than surrounding land and susceptible to flooding.

Martha Dibiossi was out sweeping. She said there had been three feet of water in the street the day before. A bike was chained to a post. There was seagrass trapped between the seat and the pole it was locked to. She lived in the Fort Sumter House, once a hotel but now harbor-view condos. It was built in the 1920s on fill. Some were cleaning out the pool filled with debris. And fish. They used pool skimmer nets rather than bare hands. She pointed to the concrete ballasts at the seawall. "Water came from over there." She knew that the city wanted to raise that wall in the city plan but worried about the "frequency and severity" of these storms. This was the second five-hundred-year (or more) storm event in as many years. The Carolinas were also hit hard by rains from Hurricane Joaquin. "Here we go again," she said, when she saw the forecast. When I asked her what to do about it, she said, "We've got to come together, face reality."

The streets were still mostly empty. Most people evacuated the area. Then governor Nikki Haley had ordered it. On my way down in the morning, people were not rushing back. The governor encouraged them to be patient. The highway lanes had been reversed the day before, southbound lanes heading north. I passed a convoy of utility trucks, cherry pickers, heading down to help with power lines. Some seven hundred thousand people were without electricity. People I spoke with said it was worse fifty miles out, in rural areas, than downtown, with a stable grid and the hospital. There was a fleet of trucks with the words *fire, water, mold, carpet*, heading down to clean up. Trees were down along the road. Branches and leaves scattered in the median. Pools of water surrounded buildings like moats. On the shoulder, armadillos that had failed at crossing the road lay splattered in the morning sun.

On my first trip to the region, things felt equally upside down. On the winter solstice, the shortest day of the year, I usually like to head for the high country to cross-country ski. But December 2015 was the warmest ever up and down the East Coast. Where I live in the southwest Virginia mountains, temperatures hovered in the sixties and seventies before Christmas, not dropping below fifty in the evening. Blacksburg, Virginia, broke the record for the high temperature (sixty-four), and we crushed the warmest low temperature of the day by ten degrees (fifty-nine). Boston was warmer than Phoenix, and the low temperature in Vermont was forty degrees above normal.

Temperatures in December 2015 were all over the place, and it was so unseasonably warm that the month broke the National Weather Service's anomaly scale. December 2015 would become one of the most anomalous months—hot or cold—ever measured in the United States. Rather than scenes of a winter wonderland, the Christmas season brought spring blooms on confused plants. Rather than head for the high country, I headed for the low.

The geographic capital of the Lowcountry, Charleston, South Carolina, is a metropolitan area knitting together peninsulas and marshy islands. Steeped in history, it played an important part in the American Revolution and was the site of the first battle of the Civil War. Today, some seven hundred thousand people call the Lowcountry home. It is also a busy sea port and popular tourist attraction, known for its charming architecture and hospitality. After South Carolina's epic storms in early October with more than twenty inches of rain in three days and historic flooding, I wanted to see if attitudes had changed regarding sea level rise and climate change. I wanted to know if the storms of 2015 were the wake-up call the state and city of Charleston needed.

Of course, it is difficult to tie the storms specifically to climate change, but the city sees more and more "rain bombs," stronger in both frequency and intensity, and average seawater levels are roughly a foot higher in Charleston than they were a century ago. That trend will continue, if not curve upward.

Charleston, like many port cities on the East Coast, was founded because it offered protection for ships from the seas, but there may be no safe harbor anymore. There were just two significant tidal flooding events per year in the 1970s compared to eleven in the early 2010s. If that continues, by midcentury, the city will flood more days of the year than not. At the end of October, after the rains, a peak astronomical tide—a spring or king tide as they call them—came in at 8.69 feet, the largest crest since Hurricane Hugo crested at 12.56 feet.

As a coastal city with elevations near sea level, Charleston has long been vulnerable to flooding. Since its founding in 1670, it expanded in size by filling in creeks and marshes. In 1837, Mayor Henry Pinckney offered a $100 gold coin to anyone who could develop a solution. One was never found. But with increasing urgency, the city tries to meet the challenge and adapt to the realities of living near water—rising water.

On the solstice, I arrived at the new city hall to meet with Carolee Williams, a project manager in the Department of Planning, Preservation, and Sustainability. Williams was fifteen minutes late, but she showed me why. She had been putting the finishing touches on the city's Sea Level Rise Strategy Plan.

In late December, Democrat Joe Riley was winding down forty years of being mayor of Charleston, longer than anyone in Charleston's history. In that time, he had seen many changes since the 1970s, including increased flooding. The devastation of Hurricane Hugo (1989) really focused Riley and the city on the issue of flooding and climate change. With his tenth term coming to an end, Riley told Williams and the task force that "I won't have done my job if we do not do something about sea level rise."

Charleston is one of the more vulnerable cities on the East Coast, sitting just seven feet above sea level. Yet, it has lacked a comprehensive plan to address a rising sea. The Century V Plan adopted in 2011 did not mention "sea rise" or "flood." In 2010, a committee released a "Green Plan." It was accepted by the city but not adopted. Hamilton Davis, Energy and Climate director for the South Carolina Coastal Conservation League, a group that played a role in the Green Plan, told me of the poor timing of the release. "It was the end of the housing bubble, the beginning of the financial crisis, and the rise of the Tea Party." Carolee Williams said there were more people than ever at the city council meeting when the Green Plan was unveiled, many wearing "green tyranny" T-shirts. But Williams and Davis are among those who see the tide, quite literally, turning.

The city has focused $235 million on underground drainage, tunnel, and storage systems. The tunnels, some of which are 10 feet in diameter and as much as 150 feet below historic homes and churches, offer storage capacity during a high tide flood and connect to pump stations that, ideally, send water into the Ashley and Cooper Rivers at a lower tide. Williams called them a "circulatory system" for the city. Laura Cabiness, city engineer and director of Public Works, said the tunnels are placed deep in the Cooper marl, or clay. Their function is mainly to collect storm water, but if saltwater gets in, it would not hurt the concrete. During the 2015 flood, Cabiness said the "system worked better than expected." Especially under the City Market, built long ago on a creek bed. The iconic tourist attraction used to flood regularly. Search the web and you will find pictures of people posing in kayaks after storms, paddling the submerged streets.

For a city their size (130,000), they are making a significant investment. Charleston's annual operating budget is about $150 million. New York City, with a budget of $79 billion, has embarked on a $20 billion resiliency plan to provide shore protection since Superstorm Sandy. As Davis noted "It took Sandy and Irene for the Northeast to get aggressive." Miami Beach, Florida, with a budget of about $300 million, has embarked on an estimated $500 million program of installing seawalls, raising roads, and installing pumps. And Norfolk, Virginia, has plans for floodgates, higher roads, and a retooled storm water system totaling more than $1 billion, about equal to the city's annual budget. "For a city our size," said Cabiness, "we're pretty active."

The new plan lays out a series of steps the city can take in the future. These include building codes to raise structures at least a foot above flood elevation, raising streets, buying frequently flooded properties, restoring wetlands, hiring a chief resiliency officer, and regularly updating their strategy to keep pace with science. They are also considering flood gauge devices, otherwise called "tide flaps." In a commentary for the *Post and Courier*, Mayor Riley wrote that "this Sea Level Rise Strategy is a comprehensive inventory of initiatives to be further vetted and costs and benefits evaluated. Looking at the risks and prioritizing the work is an essential next step."

However, the report does not address the precise location of future projects nor their specific costs. Dana Beach, executive director of the Coastal Conservation League, said the plan was an "encouraging" and "symbolic first step" but also "fundamentally inadequate" because it lacked specifics. Nor does it address any of the root causes. The phrase *climate change* is used only once in the twenty-page document. Still, Cabiness told me there are real concerns materializing about climate change. "Whether we know what's causing it or not, we are seeing climate change, and people aren't arguing anymore."

When I reached Riley by phone, he emphasized that successful communities "have to invest in infrastructure." "In America," he said "we are timid about asking people to pay more money for improvements. But if you paint a clear and vivid picture of the improvements, and the need for them, citizens accept that." To sell a plan, Riley pictures a family sitting at home, reading the paper, watching TV. What are their cares and concerns? "I have to explain to them why it is important." Then, he said, "you have to deliver." Build the project to satisfaction. Water and wastewater, Riley said, are expensive, "but you can't gamble on those pieces of infrastructure." When I asked him if climate change was ever part of the selling, he said "you have to lead and take positions that are not widely held. I don't back away from climate change as real, but I don't engage in causation if I don't have to."

Many of the experts I talked with emphasized the complexity involved in getting to where they are. Liz Fly, a coastal climate specialist with Sea Grant (a NOAA program dedicated to coastal research), said "hydrologic modeling is very complex—where exactly is the water going to flow?" She emphasized "figuring out our vulnerabilities, planning for them, and finding the right partnerships to move forward." Hamilton Davis said it can be very difficult to prioritize those vulnerabilities, even to plan fifty years out, which is much longer than most planners project. The report assumes an increase in sea levels of between 1.5 to 2.5 feet over the next fifty years, which is the mean of high and low estimates from the National Oceanic Administration and the U.S. Army Corps of Engineers. These models try to account for thermal expansion, glacial and polar ice melt, and reductions in greenhouse gas levels.

Such abstract models, far into the future, can be difficult for people to comprehend. We tend to treat the immediate and personal more urgently than the distant and uncertain. But residents are beginning to see flooding on their daily commutes, their trips to school or shopping. One person, a former chef at the Francis Marion Hotel, told me he had seen water come up through the kitchen drain.

Also, because the problem can seem so enormous, it is easy to lose a sense of personal efficacy. But social scientists are quick to point out that we are social beings who respond to social norms, so a signal from the city that they are doing something can create community-wide behavior. The city's Sea Level Rise Strategy, though still vague on specific costs and projects, does break down the problem into smaller and smaller accomplishable tasks.

On the Sunday after the storm, I walked around the downtown area a while. Many of the houses were boarded up. The historic courthouse windows were covered in a tough sail cloth. Sandbags lined the base of glass doors of banks. But the bells for church rang, calling to well-dressed parishioners. Charleston is called the Holy City because the various ethnic groups that came to Charleston brought with them numerous Protestant denominations, as well as Roman Catholicism and Judaism. It was given the name for its tolerance for religions of all types and its historic churches.

On Broad Street, I heard the bells of St. Michaels, with its giant white steeple and large Tuscan columns. Built in the 1750s, it is the oldest surviving church, welcoming President Washington himself in 1791. I poked in on the service. The homily was on the life of Paul, shipwrecked on an island, and then, stoking the fire, snake bit. Nevertheless, though he had "been in the ditch with us," he carried on through adversity. I had a pretty good idea of where the story was going and was getting impatient. I wanted to talk to people, and heard other bells.

They came from St. John the Baptist, brownstone and gothic with a tall spire, the mother church of the Catholic Diocese of Charleston. In June 2015, Pope Francis published his encyclical, saying climate change "represents one of the principal challenges facing humanity in our day." The pope says it is a moral responsibility to do something about the problem, one that has ethical and spiritual roots and will require changes in human behavior. "A rise in the sea level" will create "extremely serious situations," especially "if we consider that a quarter of the world's population lives on the coast or nearby, and that the majority of our megacities are situated in coastal areas."

I walked behind Gene Zurlow and his wife. He had been inspecting the sidewalk along the way, wondering out loud if the piles were something someone swept or random swirls of debris. They looked like the latter to me, but maybe Gene was hopeful that things were getting back to normal and people were cleaning. He reminded me of people who live in nice neighborhoods and comment on the grass. The inspector.

He told me a little about the history of the church. The secession document was read and discussed in a building that was behind us, St. Andrew's Hall, now a parking lot. The church and hall burned down before the war, and though they had Hartford Insurance (a northern company it seemed to be implied), someone didn't pay the premium. It was rebuilt with savings and donations.

Today, from what he had heard, the sandbags kept water out of the basement and all services would go on. The storm came at low tide, "so good timing." I asked Gene what he thought of the pope. He waved his hand in that manner that signals, "iffy," lukewarm, and said he's an "unusual guy." Yet, what many like about him is how normal he seems.

Inside there was a "solemn mass," which meant no incense, choir, or communion. Monsignor Brovey said it was a week of anxiety and fear, and a "storm reminds us of our mortality," that we are "not in control," *he* is. We were urged to pray for those who lost their lives, for those in Haiti, for those who protected us from the storm.

I walked on to the Circular Congregational Church (United Church of Christ), a congregation as old as the city. On a black metal fence hung a white banner that proclaimed the church's commitment to environmental issues and social justice, on another a rainbow for tolerance. The church did not have a regular service, but Kendra Luce and Jennifer Wicker had opened the nursery, in case people needed childcare during cleanup. They referred me to some of the writings of the church. After the Paris Agreement, Rev. Jim Antal, conference minister and president of the Massachusetts Conference of the United Church of Christ, stated: "The world's climate negotiators have finally realized that nature doesn't negotiate."

People were gathering outside St. Michaels, so I strode back over there, making a circle. I asked a few how they were faring after the storm. One said the storm was "another blow" but that they would pull through. Her house had flooded twice, this year and last. I asked if she thought climate change played a role. The tone of the conversation, and her body language, shifted. It was as if I had said something deeply offensive. She said it was probably an unfortunate series of storms that they happened to bear, but "I doubt it was that," referring to climate change. That the climate could be changing, disrupting normal events, was a challenge to a belief structure for many like her, that life proceeds in an orderly fashion. She gave me to know that we were discussing an area too sensitive for just after the storm, too impolitic for after church, so I thanked her and went on my way.

I thought we were talking about the weather, or more specifically climate, which is weather over time. But one reason the topic might not have been welcomed is we were a month out from election day in a grueling political season. On the day I arrived a year earlier in 2015, the only prominent Republican candidate to publicly accept the science behind climate change and sea level rise, Sen. Lindsey Graham from South Carolina, dropped out of the presidential race. Senator and former governor Mark Sanford, he of the Appalachian Trail, has seen sea level rise at his Beaufort County farm. In 2007, he penned this for the *Washington Post*: "For the past twenty years, I have seen the ever-so-gradual effects of rising sea levels at our farm on the South Carolina coast. I've had to watch once-thriving pine trees die in that fragile zone between uplands and salt marshes." However, Republican congressman Bob Inglis was defeated in 2010 for acknowledging the realities of climate change. Inglis now runs an organization to persuade conservatives that there are ideologically pure ways to deal with global warming. He received the John F. Kennedy Profile in Courage Award for public servants who take principled but unpopular stances. With a 100 percent rating from the Christian Coalition, he hopes to "help fellow conservatives to see in climate change a moon shot opportunity for more energy, more mobility and more freedom." Katherine Hayhoe, a climate scientist at Texas Tech University and outspoken Christian, has said that immigration, the economy, national security—all conservative causes—relate to climate change.

I reached former congressman Bob Inglis one day on the phone when he was heading back from talking to a Boy's State group. Boy's State is put on by the American Legion and skews conservative, founded to "counter the socialism-inspired Young Pioneer Camps," according to the American Legion website. One of the things I wanted to ask him about is how he talks about climate change to resistant groups. First, I wanted to find out how he made the turn he did, toward bucking the conventional wisdom (or lack of it) in his party.

For many years, he thought climate change was nonsense. He represented the Fourth district in South Carolina (Greenville and Spartanburg), from 1992 to 1998. He took a six-year hiatus but ran again in 2004. His son, who was then voting age, said to him that "you're going to clean up your act on the environment." By that time, his family had moved to a "farmette" because Inglis wanted rivers and open space for his children to play in, "creeks and critters."

His participation on the Science Committee in Congress also influenced him. As part of that work, he traveled to Antarctica to examine ice core drillings. The South Pole is a desert, receiving only a quarter of an inch of precipitation a year. In those ice cores, scientists can find a record of the amount of carbon dioxide over time in the atmosphere. Seeing the evidence up close, it started to make sense. "We affected the chemistry of the air. Light comes in, but heat gets trapped."

The third thing he described as a kind of spiritual awakening happened while snorkeling with scientist Scott Heron at the Great Barrier Reef. Among the vivid fish and coral branches, Inglis realized that Heron, too, was worshiping the spirit as he interacted with nature. "Scott would show me something, come to the surface, and shout, 'This is amazing!'" Inglis realized that science and religion could coexist and that both might be useful against the threat of climate change. He told *Frontline* that "God wants us and science to discover this creation, so why wouldn't we listen to the people who have given their lives to this endeavor, who have learned and who know things that I don't know?" Why instead, he continued, do we "decide to listen to some people of not-so-credible backgrounds" who have been funded by efforts to introduce doubt where very little exists?

Katherine Hayhoe, the Texas Tech "climate evangelist," has said that "having a relationship with god is one of the most incredible experiences that a person can have." However, the hardest part of being a Christian is the amount of disinformation targeted to the faith community.

Inglis came home from Australia and introduced the Raise Wages, Cut Carbon Act of 2009, "not the best idea in the midst of the Great Recession, in perhaps the reddest district in the reddest state in the nation."

He told me "some other heresies got me into trouble," such as a vote for the Troubled Asset Relief Program (TARP). He also voted against a troop surge. In the "midst of crisis," Inglis said, "orthodoxy becomes paramount. The tribe is under pressure, so the tribe must close ranks." At a big tent rally during the primary in Spartanburg, a local conservative radio host asked both candidates if climate change was real and human caused. Inglis said yes. The crowd hissed. His opponent, Trey Gowdy, gave a political answer. "Inasmuch as it hasn't been proven to the satisfaction of the people that I represent, the answer is no, there is no human causation in climate change." Though Inglis had a 93 percent approval rating from the American Conservative Union, 100 percent from both the Christian Coalition of America and the National Right to Life, and an A from the NRA, Gowdy whipped him, taking almost three-quarters of the conservative vote.

In Australia, he was very happy to hear about the faith component in climate change. He said in the United States the "language of the left can be very disrespectful of a faith worldview." He said some are equally "evangelistic about godless evolution" as many in the religious community. He told me of one gathering of scientists he attended, and the speaker asked people who believed in evolution to raise their hand. Inglis wanted to half-raise his.

It was like a reverse kind of "Salvation," the Langston Hughes story where the narrator has to say he is saved when he is never really sure if he is.

When Inglis communicates with audiences, he tries to understand their values. Communitarian, egalitarian values are a turnoff to conservative audiences. He also speaks to stories as much as science. In the book *Don't Be Such a Scientist*, Randy Olsen urges people communicating science to lead with emotion, tell a story, captivate the imagination, and finish positively.

Inglis brings a story and positive message, of a kind of "high octane conservativism," of free enterprise, and of a transparent, accountable marketplace. His nonprofit, RepublicEn.org, based at George Mason University, advocates for the simple pricing of carbon dioxide. He wants to eliminate subsidies for electric cars and solar and wind energy, a popular position among the conservative faithful, but also those for drilling on public land and the ability to depreciate oil and gas expenses. He would even like to see gas priced at a rate that reflects national security. Doing so would unleash innovation.

Also, he wants to get rid of the "biggest subsidy of them all," that we can "use the sky as a trash dump without paying for the damage we're doing there." I said the last part sounded communitarian, as we share the air, and he acknowledged the "last parts are not gospel," but people need to understand the way climate change will cause havoc. It might be better called climate havoc, or climate disruption, than change.

When asked about efforts in his home state, Inglis said he was proud of the third-party financing of rooftop solar, which had statehouse support. But he still has a long way to go to reach the unconverted. When he showed NOAA inundation maps to his siblings, who live in the Lowcountry, one brother reacted strongly and did not want to see them. When people feel like they are stuck, some turn away. Denial may be a way of coping, an adaptive strategy. "It must not be real." And faith is an alternative pathway, one of belief. Psychologist Thomas Plante, editor of a book on spirituality, mediation, and health, says, "spiritual practices can be a foil to anxiety and depression."

Seeking more answers to why some in the religious community turn away from climate change, I picked up *Between God and Green: How Evangelicals Are Cultivating a Middle Ground on Climate Change* (2012). Writer Katharine Wilkinson examines what she calls climate care, a faith-based approach to addressing climate change. Wilkinson and other experts have cited an association with paganism as one reason some evangelicals turn away from environmentalism. Another reason is the "biospheric egalitarianism" of deep ecology, where creatures are on equal footing, not the Christian perspective that sees humans as unique among beings. Others, such as journalist Bill Moyers have asked "why care about climate change when you and yours will be rescued in the rapture?" God will take care of things, and us.

In Wilkinson's book, she surveys a previous study that examines evangelicals' views of race to suggest that church-based discussions also work against a systemic understanding of the causes and possible solutions of social issues, including climate change. Deeply embedded theological notions that date back to the Reformation, such as "accountable free will individualism," suggest individuals act independently of structures and are personally accountable for their actions. As the individual is the site of religious experience, so are they reluctant to cede control for personal or any other kind of salvation.

Wilkinson talked with Kevin Wilson, an evangelical pastor and writer who told her he thought evangelicals could go for adaptation to move beyond scientific skepticism and political disagreement. In her words, "adaptation might be a roundabout path to accepting climate science and mitigation measures and to more extensive evangelical engagement with climate change." Wilson and Tri Robinson coauthored a plan of action for mobilizing evangelicals who would "aid in adaptation efforts, especially through global missions, where evangelicals are already helping populations vulnerable to the dangers of a warming climate." Adaptation resonates with evangelicals' more "established history of engagement in relief and development work."

For his part, Inglis worried about the effect of the current administration on young conservatives, like his son, and those at Boy's State. Opinion makers matter, especially in a faith community. He feared that leaving the Paris Agreement could have a negative effect on young conservatives and try their faith in the party.

Later in the afternoon, after my fish rescue and church attendance, I drove out to Sullivan's Island, a barrier island, to check out what the storm had done to the beach and to nearby homes. Both it and the nearby Isle of

Palms, an island just north of Sullivan's, were hit hard in the October 2015 rains. Much of the dunes in what locals call the IOP were eroded. Million-dollar homes had to put up sandbags. Hamilton Davis, from the Coastal Conservation League, told me about places there that are a "visual illustration of what sea level rise will mean." We have to "armor our beaches to protect properties, but then we lose the resource."

When I got to the island, police were stopping nonresidents on the island side of the causeway, to prevent looting I supposed, which only made me more determined to check out the scene. I made as if I was turning around in the supermarket parking lot. And when there, dipped inside for water and a snack, as if that would somehow help cover my tracks or make me seem more local. Inside the market, there was no water left, which seemed logical. If people had no power, they would want clean water. But there was also no toilet paper on the shelf, which did not seem to make obvious sense. Only that it was some compulsion to stockpile, to feel in control during the coming chaos. The only sandwiches left were pimento and cheese. I felt like one of the seagulls, picking over scraps.

It is an odd experience to walk to the beach in the Lowcountry. You can see the ocean but you seem to be at eye level with it. I walked out on a boardwalk and over about one hundred yards of grasses and marsh, still some standing water, dunes a storm surge would seem to laugh at. Indeed, much of the area and the front yards of large beach front homes, all raised on pilings, was under water in October 2015. Not as bad in 2016.

On the beach, a few clumps of spartina grass had washed ashore. Few people roamed with me. The sand had been washed over, leaving ripples and no footprints, hard under foot. There were root systems to plants I could not identify, gangly and alien. The surge had scarped a portion of the beach, snaking a new channel out to the sea.

I talked to someone on the beach, putting away his kiteboard. It has a hydrofoil on it that allows him to skim across the surface of the ocean, giving the sensation of flying. Like the nearby houses, the theme in the Lowcountry seems to be elevation above water. He owns an adventure sports shop on the main street that has flooded three times in the past year.

Folly Beach, across the harbor and south of the city, got hit even harder. "Where there used to be sand dunes is now rock," Mayor Tim Goldwin told the *Post and Courier*. And though storm damage was moderate to property, to the beach it was extensive. So much that sixteen Civil War cannonballs were exposed. The local sheriff had to call in a bomb squad.

In 2014, the U.S. Army Corps of Engineers completed a $30 million beach project there. Hurricane Matthew removed most of it, and the town will likely want help again. As sea levels rise and climate change intensifies storms, pumping sand onto these beaches can seem like folly. We will eventually run out of sand. Or money. Or both. There may indeed be places where nourishment projects are needed for coastal defense, or for economic reasons, but there seems to be no national discussion on where those places are or which are a priority.

On a previous visit, I had a chance to visit St. Helena Island near Beaufort. Both Hamilton Davis and Liz Fly had told me about some of the good things the mayor of Beaufort had been doing, including establishing a sea level rise task force. When there, I visited the iconic house in the movie *The Big Chill*. Like many properties in the area, it experiences more and more flooding. The railing on the long dock that leads into the bay, the one visible from the kitchen window when the cast dances while doing dishes, had fallen over. Fiddler crabs skirted in and out of holes on the front lawn. I called the realtor who once had it listed for $4.5 million, inquiring about flooding. She told me it did not flood while she had listed it, which seemed a roundabout way to avoid full disclosure.

On that first visit, I joined Queen Quet and the Gullah/Holiday celebration at MJs Soul Food Sea Island Parkway. I found the queen at the book table. She had just returned from a conference in Boston, yet another where she was asked to talk about sea level rise and relocation. "We are not vanishing!" she told them, and me, and I was reminded of the convenient narrative of settlers of all kinds that natives were vanishing when such a view merely excused their removal.

Marqueta Goodwine was selected by her people as spokesperson and head-of-state for the Gullah/Geechee Nation, and she has represented them around the world, including at the United Nations. Keeping her culture alive is her main goal. The first Gullah/Geechee people brought to the Carolinas were blacksmiths from Angola. *Gullah* may be a corruption of *Angola*. *Geechee* is likely a corruption of the Ogeechee River near Savannah, Georgia. Their skills and tools were also used to clear the land for plantations that grew cotton, indigo, and Carolina Gold rice. According to Queen Quet, the enslaved Africans sold in the nearby Charleston market were called "black gold." Because of their isolation in rural areas, the Gullah developed a culture that preserved much of the African linguistic and cultural heritage as well as absorbed new influences from the region. Escaped slaves often headed for the marshy isles. In Toni Morrison's *Beloved*, Paul D and the prisoners he is chained to, all forty-five of them, would have "headed for the Sea Islands that slid down from the Blue Ridge Mountains. But they didn't know."

Sooner or later, conversations about sea level rise and adaptation turn to relocation and retreat. In one recent discussion Queen was invited to participate in, she told me she was struck by the differences in her approach and others at the table. "The values they were focused on were monetary and the values I was focused on were cultural." "I have a degree in mathematics," she told me, her gold earrings flashing along with a smile both serious and sassy. She wears her hair in braids, tucked under a crown of cowry shells. "I can do cost assessments. But people aren't buildings. They aren't something to be managed or engineered."

I caught up with her again at a Coastal Cultures Conference, "gwine ta de wata." The queen talked about the possessiveness of people and how developers saw land as object: "This is a paradise. So we want this." But people with a long history in the place could see that building where they wanted to build was not a good idea. There was a reason her people did not build there. She spoke of surviving both colonization and enslavement. She said that her ancestors have always had to adapt, coming from the western seaboard of Africa to the eastern seaboard here, from Jacksonville, North Carolina, to Jacksonville, Florida, surviving hurricanes and tourism and overdevelopment in places not ecologically sound. "For us, the waterways are our bloodline and the land part of our family. So we never want to harm it." "We live on the water, the tide shifts and changes, and we shift and change like water." The words *adaptation* and *resiliency* do not exist in their Creole, English-based language, "but we are a resilient people." "We teach our children to get things from the water only when they are in season. We don't cut down our trees and try to keep a buffer. We put oyster shells back in the creeks, rebuilding reefs, so the spartina grass continues to grow," holding back erosion. "What others call adapting, we call living, living our traditions as our elders did." They see less erosion in part because they develop less than other nearby islands, such as Hilton Head. Nor do they develop where they shouldn't. "A respect for our waterways is cultural."

Outside, I spoke with Ricky Wright, the vice president of the Gullah/Geechee Fishing Association, ladling fish out of the fryer and into the pan. He told me he used to catch much more in the creeks, but regulations have gotten stricter, have taken the fun out of fishing. Too, they favor commercial over subsistence fishermen, where you are limited on what you can catch in the creek but can go to the store and buy the same. He thought pollution was a part of the decrease in available fish, especially from the golf courses. At the conference were presentations about why people in the Gullah community had a higher rate than the general population of lupus. They thought it had something to do with the concentrations of toxins in a fish-heavy diet.

Quet told me that the purpose of the conference was educational, and I was struck by her faith in the future. "We seek to be sources of education for others so that they will see why our lifestyle is an extremely compatible and sustainable one for this coast." She spoke of traditions and cultural ties so deep no storm could take them away.

Walking around Charleston, I noticed a mural on a historical building on Bay Street. The mural is diptych, but the street and buildings are the same. In the left frame, things are pretty normal: people fill the street, though an alligator pulls someone in a cart. In the right frame, water has flooded the cobblestone streets, right up to the buildings. A woman sits on a dock with her toes in the water, and a man tends the grill. People lounge on a covered deck, and a monorail emerges from behind one of the buildings, most of them covered in more vegetation. The artist, David Boatwright, told me it was a "whimsical piece" that allowed him to put in some friends that have passed on, including Edwin Gardiner, a bike advocate killed riding a bike after a morning row on the Ashley River. Gardiner is in both paintings because "if anyone could transcend time in the story of Charleston, it'd be him." In the future-looking painting, "the streets are flooded," said Boatwright, but "Charlestonians adapt."

The night after the storm, I stayed with an old friend who lives near the Colonial Lake area, a couple blocks in from the Ashley River, on the western side of the peninsula. He evacuated the night before and watched on the evening news a truck attempt to drive down his street. Water was up to the doors. It was now parked in the lot across the street, likely immobile.

The peninsula is a kind of oval bowl with a lip around the rim. The lip, or seawall, is uneven. Some of the lip has been raised only moderately above the typical high tide to five or six feet. City engineers talk of raising the High Battery seawall, which protects the tip to five or more feet above a seven-foot king tide. They want to build up the Lower Battery seawall to account for another two and a half feet of sea level rise. At a cost of $48 million, generated through hotel taxes, the Lower Battery project will take six to eight years to complete, though it will still offer little protection for sections beyond its boundaries.

It would have likely helped my friend Pat's neighborhood. When he arrived, we peeled away garbage bags he had duct-taped across the door and threshold. The inside lower landing had some moisture damage, but his first floor seemed fine. In the patio area though some bottles and trash had washed in. He had created a fish pond back there. The fish had been washed out, maybe joined with the morning's release.

We ordered pizza and settled in. He called his ex-wife, Louise, and their son to check in on them. I could overhear parts of their conversation. Louise lived a couple blocks from the Ashley River in an old historic 1875 home. She flooded last October and again this time. The hardwood flooring was warped, the dishwasher likely ruined. "I can't live like this," she told Pat, over the phone. "I have to do something."

She was thinking of either selling or raising. But raising is complicated in a city that likes to preserve the historic character of its buildings. I reached out to both sustainability manager Williams and city engineer Cabiness to see what they knew about helping Louise. Cabiness gave me the name of a person in the city's Design and Preservation Division, who was willing to visit the house and talk about raising and guiding the process through the Board of Architectural Review. They gave me the name of a council member who could help.

Louise's home again flooded after Hurricane Irma, which tracked more directly into Florida and Georgia but whose wide, whirling bands lashed the city with rain, flooding it once again. She was by now determined to raise it to nine feet above sea level. She was also looking at something like a kind of tile for the first floor, almost like that of a boat (or shower), where water come in but does not affect the structure. After a flood, the debris can be hosed out.

And she was still looking to the possibility of a "repetitive loss" program. The city puts in one-quarter of the value of the home and Federal Emergency Management Agency (FEMA) pays the rest. However, I learned later that they did not receive the grant money for this, in part because they already received money for a previous year's flood. However, Charleston has done much to get in FEMA's good graces. Municipalities earn points for public information, mapping, regulations, and other flood preparation. They have a score of six. Cabiness said they were shooting for a five. The lower the rating, the higher the discount on insurance.

Walking around the city and talking to residents, I was struck by a sense of civic pride. A woman at the visitors' center told me at least a quarter of Charleston's population had moved there from somewhere else, about forty people each day. There was a vibrancy about it, even after the storm. And people wanted to help each other. Many days later, I thought of Boatwright's mural, of Gene Zurlow wondering about piles of palms on the walk to mass, and Martha Dibiossi sweeping up trash by the battery.

At one corner, I saw a sign for the Preservation Society of Charleston that says "gut fish not houses." I wondered if the historic nature of the city, the quaint architecture and manners, impeded progress on sea level rise and climate change. But the other preservation society in town, the Historic Charleston Foundation, had recently hosted a lecture on sea level rise by Doug Marcy, coastal hazards specialist for NOAA's Office for Coastal Management. Marcy helped advise the city on their planning projections for the report. It seemed their interest in preservation motivated not hindered efforts to do something about climate change.

While waiting for Pat, I walked in a perimeter around his place. On one waterlogged street, TV cameras had set up lights. A woman in a yellow polo and rubber boots prepared to be streamed in to the Weather Channel. Only I did not know who it was. She gave me her name, Maria LaRosa, a well-known personality, but looked at me like I might be kidding when I asked.

In my visits to Charleston, I never did run into their resident celebrity and part owner of the minor league RiverDogs, Bill Murray. In one of his most familiar roles, the weatherman Phil in *Groundhog Day*, Murray finds himself in a time warp, repeating the past. The flooding Charleston keeps experiencing can seem like that. But about halfway through the movie, Murray's character realizes that only by becoming better, improving, can he break free from the loop.

In a picture posted online, Murray stands in his pajamas, hair disheveled, seeming to reprise his role from *What about Bob?*, holding a sign upside down that says, "Mr. President. Action on climate change NOW." The caption reads, "Climate change is important because it turns the world upside down!"

December 2015 seemed like an upside-down month, and flooding in three successive Octobers has tried the patience of people like Louise. And yet things are also coming into alignment. With all the efforts going on in the city of Charleston, it seems like they have crested a very important, if low-lying, hill. To get out of the flooding time warp, Charleston and the Lowcountry may not be able to hold back all the water, but they are learning, like Phil the weatherman, and stepping out of the past.

5

Ebb-tide Optimism

Ghosts of the Golden Isles, Georgia

As I piloted the car through the misty gap in the Blue Ridge escarpment and down into the coastal plains, it occurred to me that that this was a journey made before me. I myself had made it many times as had many traveling I-77 down from Ontario and Ohio to Charlotte and points south. But in the dense fog that often clouds this break in the mountain, called Fancy Gap, I thought of the journey rock and rubble made long ago. These Appalachian Mountains were once as high as the Himalayas geologists think, but they have been gradually eroded by wind and water. They've dwindled away because they had the time, being among the oldest anywhere. Over eons, their knife edge has become rounder, though a few bony plates remain on the ridges. Large rocks became smaller cobble, became pebble, washed down the creeks and rivers until eventually pounded into sand and deposited on some far shore.

This section of road is often socked in by a dense haze, once causing a seventy-five-car pileup. Above the cloud was the faint outline of the green mountain, but above that was a darkening veil, as if the shadow of the former mountain, its ghost.

We were headed for Savannah, known for ghosts, and we followed directions on a phone's GPS until the device went out, departed. I turned down a narrow side street, which seemed close to our turn, the canopy overhead full with live oaks, snaky, dramatic limbs leaking Spanish moss. As if stunned by the view, I stopped in the middle of the street, hoping not to cause a pileup behind. My wife stepped out to snap a picture until she remembered the phone was dead. I looked at the number of the house we stopped in front of, by some chance, our Airbnb lodging for the night.

It wouldn't be the first time in my visits to Georgia that I was stopped short by trees. Or unusual occurrences.

My first meeting was on Skidaway Island at the Marine Education Center. The entrance features native plants, southern wax myrtle, beautyberry, St. John's Wort. There are more live oaks dripping with Spanish moss, which is a relative of the pineapple family. Yellow-bellied sliders, a native turtle, caught some sun on a log but dove under, disappearing when I approached.

Inside, I was greeted by John Crawford, an older man wearing a white T-shirt, shorts, and water shoes. "We wear them in the field, wash 'em off, go to weddings." I rarely feel overdressed. Certainly not today.

John earned the nickname "Crawfish" because he was always catching stuff at recess when he was younger and bringing it back to class. "What do you have, a toad?" someone said about the lump in his pocket. On that day it was a dead bat. Like many nicknames, it was meant as a kind of insult, but he took to it.

He took me to a map near the aquarium, a nautical chart, and started giving an orientation to the Georgia sea coast. The island we were on is Skidaway, some forty thousand years old, formed during the Pleistocene epoch.

Sea level has fluctuated, and out on the sea floor, experts can see the scouring of icebergs. The last great continental ice sheet froze enough water to lower the sea level as much as five hundred feet, placing the shore out by the continental shelf, some seventy-five miles offshore. At one time, the building we were in would have been beachfront, when the sea was six feet higher. He pointed to Wassaw, east of Skidaway, which dates back four to five thousand years ago. There are basically two kinds of barrier islands. Those like Skidaway, closer to the mainland (Pleistocene), and those like Wassaw, formed more recently, at the end of the melting of that ice sheet, Holocene.

Things have happened here. The flicker of the past is always present. Things are still happening.

He traced a kind of north–south line that runs through Skidaway, Ossabaw, St. Catherine's, Sapelo, a former shoreline. There are eight islands and island groups along the hundred-mile coast from between Savannah and the St. Marys River on its way to Florida: Tybee, Wassaw, Ossabaw, St. Catherine's, Sapelo/Blackbeard, St. Simons/Sea Island/Little St. Simons, Jekyll, and Cumberland. Of these, only Jekyll, St. Simons/Sea Island, and

Tybee are accessible by road. Because of the lack of access, only these three are developed. Most have little population density. They are called the Golden Isles either because the marsh grass turns a shade of amber in autumn or because of a particular tonal quality of the setting sunlight when it gilds the sand and marsh.

Crawfish aimed at the Ogeechee River, which he said has cypress trees one hundred years old, now dead or dying, the salt water moving some three to four miles upstream. At a bend in the river, he told me "they used to grow rice, but it's now marsh. Marsh is taking over."

We walked back to his office, past the touch tanks with turtles and other sea life. Georgia has a curve in it, a kind of eddy in the ocean's current. On a map, Brunswick lines up longitudinally with Pittsburgh, Savannah with Buffalo. Tides here are normally seven to eight feet but can range up to ten to twelve feet, the highest south of Cape Cod and north of Argentina. The geologic and ecological development of the coast is influenced by these tides. The concave shape of the coast is called the Georgia bight, and it may be one reason why Georgia had not experienced a major hurricane until Matthew in 2016 (the first since 1893).

"I've seen it happen in my lifetime," he said, referring to climate change. Crawfish called up images on his computer, places he would visit by boat, little hammocks or islands of trees that are now skeletal, their spiky forms still visible by satellite. To demonstrate how those islands formed, he shook a jar of red clay, feldspar, and sand, which is quartz. The sand settled to the bottom then the silt layer, with clay on top of it, and finally organic material resting on top. With rising seas, that process reverses. He had a tube of marine epoxy on his desk and a picture of knots hung on the wall: reef knot, anchor bend, bowline, half-hitches.

On Google Earth, he highlighted the shadow of a former river, to low lying areas now inundated, to freshwater ponds now briny. He studies frogs and amphibians, and many of them have been displaced. I called the area we are looking at a wetland but he didn't like that term, "lumps too much together": gum slough, cypress pad, ox bow lake, pine flatwood, shrub bog, cypress dome. They are even called batteries in Okefenokee, each a different biological community.

He scrolled images through the screen, waving through islands and woods, some "wrecking" over here, a shipwreck faintly visible, then a shell midden on one island, from four thousand years ago, the shape of a horseshoe. Indians may have hidden behind it during floods, a protective ring. They were a community nexus, a place of public interaction, and might have helped adapt to environmental stresses such as storms or mosquitoes.

If they built structures, they were on the back side of the island, as were the Spanish missions, nearly opposite to building patterns now. He shook his head at something, "there's no way to win," thinking about rising seas and what he had seen in his lifetime. "We shouldn't build anymore on outer islands," he said.

"But we have," I said.

"It's only a matter of time," he said, shaking his head, adding "Hilton Head," and making a face, tightening neck muscles, like he had seen a ghost.

Outside, he showed me where the tide came over the seawall last fall, marsh grasses from that event still visible, some signs of erosion.

While on Skidaway, I also paid a visit to Clark Alexander, an ocean geologist with the Skidaway Institute of Oceanography. He had been there for twenty-six years and had never seen water on the dock until that past fall. Clark is a serious and respected scientist. Many of the folks I spoke with deferred to him. When I contacted Alexander by email, he asked pointedly, "Why is someone in English working on climate change?" I did my best to answer, saying how I wanted to move it out of abstraction, to bring it down to a local level in ways that real people experience it. He consented to our meeting.

I have credentials in writing and literature, but in the presence of some scientists, I worry that I'm perceived as ethereal, a frill. That was my experience working as a technical writer. And too often, scientists are perceived by some I work with as being pens-in-the-pockets detached, hyperobjective, lacking the larger context and story. It can be hard, in my experience, to move scientists over toward the policy implications of climate change, which borders on "activism." He agreed that experience makes people more accepting. "They have the same reality. The same set of facts."

I could see he was busy when I arrived and worried that I was wasting his time. I had questions about what we were seeing and how we know what we see is climate change. And I asked about past and present levels of sea rise and what metric he knows of, besides tide gauges, to show a rising sea.

A student interrupted us. He had a question he wanted clarified, but it was one he might also have known the answer to, though wanted to check, making that careful calculation many of us have made with a boss, especially a demanding one: If I ask, will I seem stupid? Then again, if I don't, will I make a mistake? He asked, but, since he interrupted, was dispensed with curtly.

I felt honored that my time was given such importance. When Alexander resumed, he talked some of how difficult it could be to know about changes. Even some of the evidence in the dunes of past levels of sea rise might have been the result of previous dredging. He mentioned a term, the *tidal prism*, for the volume of water that fills the back-barrier environment every tidal cycle. As sea levels rise, the tidal prism increases, leading to natural erosion of channel banks. But boat wakes and wind-created waves are other sources of channel erosion, so it can be difficult to know precisely what is causing the erosion.

Before I left, Alexander offered an anecdote. He was speaking once to a planning commission about what he saw in terms of sea level rise, and he talked about the marsh and salt water moving further inland. He spoke about how that line is moving inland as the sea rises, and the effect on the freshwater species, including trees. One person stopped. A light had gone off. "I've seen that in my lifetime."

As a result of the work of Eugene Odum, often called the father of ecology, Georgia passed the Coastal Marshlands Act of 1970. Georgia does not allow building below the high-water line. (Some states, like Virginia, actually allow building, such as docks or other structures, to the low water mark.) This has helped slow development along the Georgia coast. Odum proposed the outwelling hypothesis, by which healthy coastal marshes produce an excess amount of carbon each year and "outwell" these organic nutrients into the surrounding coastal ocean, increasing the health of local fisheries and ecosystems: protect the estuary to protect the fishery. The theory is hotly debated. But in Alexander's words, Odum helped convince people that the marshes were something more than "bug-infested swamps."

That evening, I drove out to Tybee Island, a beach community near Savanah. The island is about three miles long and a half-mile wide. I turned around several times in the small residential housing area where I was to meet my next contact and was beginning to think I was lost. Maybe it was one of those streets with opposite ends, a north and a south. Though it seemed like I was there from the way he described it, and he gave directions like people used to, "go over the bridge and you'll see a store and then the road bends right. Take a left there by the water tower." I was to write this down so I would remember, but I knew I had his address, so was led around by phone, which said the number five should have been right where we were, but there was no five, only four and a six. I decided to knock on the one in-between and just as I was about to, a face appeared from behind the curtain, a shock of white hair. It was not a ghost. It was Paul Wolff, and he was finishing putting up the five above the garage. After we shook hands, I helped him eyeball center the number, so future guests would know where they were.

I followed Paul and his bare feet as we climbed up a spiral staircase to a rooftop deck of the condo, his girlfriend's, not yet home from her work as a nurse. From there, we had a view of the Savannah River and across it, Daufuskie Island, where Paul has a cabin.

I wanted to talk to Paul because, as a city council member, he was responsible for the adaptation plan the city of Tybee produced.

It started with a "surfer girl," Ashley, who approached him about limiting the use of plastic bags. Wolff went to Vanderbilt in the 1960s and joined the back-to-the-land movement, raising organic cattle before there was organic. He moved to Tybee in 1994, falling in love with the place on a vacation. He wanted to make a difference in it so he ran for council and started a number of initiatives. When Ashley approached him about the bag ban, it seemed a natural fit. Those bags are wasteful, they are made from oil, and they often get swept up by the coastal winds and end up trapped in trees, like ghosts. They litter sidewalks, streets, and beaches, or they are carried into the ocean by storm water drains, where they will never biodegrade but where the chemicals in them will break down, releasing toxic endocrine disruptors into the environment. In the ocean, many animals confuse the bags for jellyfish. One in three leatherback sea turtles have plastic in their stomach, most often a plastic bag.

Los Angeles banned them in 2013 by a vote of 11 to 1. San Francisco dropped them in 2012. Eventually the whole state of California banned single-use plastic bags. Tybee could keep with the times, be like progressive Athens, Georgia.

Then, he started hearing rumors: about how reused bags might carry salmonella, about how older people could not use reusable bags because they hurt their hands. When the issue came up for a vote, the Koch brothers sent in lobbyists. Soon, there was a bill banning bag banning moving through the legislature. Senate Bill 139 prohibited communities from restricting "single use containers." Suddenly, the Tea Party and the Kochs were on opposite sides because the Tea Party did not want to cede home rule but the Kochs wanted to nip this effort in the bud, under the banner of "consumer choice," as they have made their fortune on petroleum processing. The Tybee Market opposed the ban but admitted that people had started using them less because of the awareness the issue generated. "What about the tourists?" people worried. Riverkeepers and the Sierra Club kicked in some money for free reusable bags. They could be souvenirs, something to take back home with Tybee written on them. The initiative to restrict plastic bags lost by one vote. When Wolff came up for reelection, he lost by ten votes.

In what he calls his "aborted political career," Wolff put the city on the front line of sea level rise adaptation. Their report, which the city manager gave to me, helped raise awareness about the city's vulnerabilities. Because of existing efforts in water conservation, storm water management, and the preservation of open space, along with communication about vulnerabilities, the city earned "credits," moving them up two classes in FEMA's Community Rating System, from a seven to a five, tied for the best rating in Georgia, though they are on the coast. The rating system seems a little like insurance karma: do good things, and you will be rewarded. Tybee's flood insurance discount went from 15 percent to 25 percent, saving some $3 million for property owners.

And then there's the road.

Tybee is a significant driver of the state's coastal economy, a tourism hub, but to get there, you have to travel a narrow causeway out to the island, US Highway 80, the only road on or off. The mayor, Jason Buelterman, told me it floods a dozen times a year, but in 2015, it was twice that (some of those were at night, when not as noticeable). By 2050, it could flood forty to fifty times a year. On the way in, I noticed the marsh grasses washed up on the shoulder. Right near the entrance to Tybee is Fort Pulaski and a tide gauge. Three of the top ten highest tides ever recorded occurred in October 2015, flooding the road, cutting it off from the mainland, a trend that will continue. Jason

Evans, a professor of environmental sciences at Stetson University and the author of the adaptation plan, said he often shows audiences a picture of the road when the tide was 10.2 feet. It floods at 9.2, so it was a foot underwater, impassable. When underwater, helicopters are on the ready. The coast guard is called in case of emergency.

Mayor Buelterman told me they have had plans to do something about the road for twenty-five years, but in the last two years, they have come closer. At one time, the Corps of Engineers wanted to widen the road to four lanes but not raise it significantly. Wolff and others worried this would change the character of Tybee, which he told me is referred to as the "Redneck Riviera," though he preferred "Mayberry on Acid." Wolff called it the "monster road." Raising it is a complicated feat as there are many factors: wetlands, a National Park (the fort), and endangered species (terrapin). Raise it and you can alter those habitats, adding fill, cutting off natural flows. According to the mayor, to raise it about a foot and half to two feet will likely cost about $110 million. It is a visible indicator of the challenges of adaptation to climate change.

I had a hard time scheduling a meeting with Mayor Buelterman, so I wondered if he was blowing me off. As US 80 is such a visible indicator of sea level rise, the mayor is contacted often. "I'm a Republican," he told Justin Gillis of the *New York Times*, "but I also realize, by any objective analysis, the sea level is rising." Buelterman said he has tried "everything short of a hunger strike" to get a new road. His argument used to be based on economics but now it is based on vulnerability.

I ran into him at lunch—we were eating at the same restaurant, the Green Truck Pub, and he agreed to talk more by phone. He told me of other things the city had done, such as capping freshwater wells. He told me they worked on homes in "velocity zones," vulnerable to waves and flooding. They built up dunes, protected them, and put in crossovers for access. After the 2016 Hurricane Matthew, the mayor could be heard on NPR crediting the dunes with saving houses. "I'm breathing a sigh of relief from what I saw," he told reporter Rae Ellen Bichell. "It's unbelievable that I'm standing right here and not seeing massive damage."

The hurricane was the worst to hit the Georgia coast in over a hundred years. A tidal gauge at Fort Pulaski, just outside town, hit a record twelve and half feet. Flooding typically starts at ten. The mayor praised the vast majority of Tybee residents who heeded his advice to get out of Dodge. Paul Wolff was not among them. He told me the walls shook in the ninety mile per hour winds. And there were trees falling, limbs breaking, loud snaps and cracks outside. No power. Just the dark, scary night. "I lay awake, safe behind my storm shutters, listening to things going bump in the night, wrapping myself around a vodka martini." I asked him if he would do it again. "I would consider carefully."

I did meet with city manager Diane Schleicher at city hall. Outside the building is a post, like one that measures snow, only this one for the category of hurricanes. With a Cat 3 hurricane, water would be at my waist. In a Cat 5, it was over my head, water to the first-floor roof. We talked in the auditorium where meetings are held, banners that read "Site of Olympic Volleyball, 1996" during the Atlanta games. Only the beach volleyball never happened. Officials were worried about the road.

Diane told me about further strategies for when it floods, such as a reverse 9-1-1 system to let people know. But she was unsure about climate change, "You don't know if you're going through a cycle or not."

On another trip to the Georgia coast, I took a naturalist tour of Little St. Simons Island. I went in part because I had heard the organic garden there had flooded several times. Hank Paulson, former Treasury secretary under George W. Bush, purchased the island and former hunting and now ecotourism lodge along with his wife, Wendy. They put the 11,333-acre barrier island under a conservation easement and seem to be good stewards of it, but it seemed a delicious story that, whatever the ravages of climate change, the cucumbers of the man who many associated with the financial collapse of 2008 were a soggy, sodden mess.

What better way to show sea level rise? However, naturalist Stacia Hendricks told me that while the garden was indeed prone to flooding, it was not very old, so a poor metric for climate change. And besides, the greens and peppers I sampled at lunch were terrific.

With several other travelers, we rode around the sandy island road in seats in the back of a pick-up. The island is a biological treasure. Our driver pointed out things to see: painted bunting, stunning color, my first; glossy ibises; and roseate spoonbills. Like flamingos, they take on the color of their diet—in this case, the pink in the carotenoids of shrimp and other shellfish.

We watched a few terns dive-bomb their prey. I had hopes it was an arctic tern, which migrates from one pole to the other, some kind of climate change messenger thrown off its normal route, but the naturalist told me to keep trying. We wondered at the movement of something in the channel at the end of the island, some larger species stirring up a school of fish so that they roiled the currents while climbing over and around one another, but it never surfaced. Some mysteries stay hidden from view.

Many people I talked to said that if I wanted to learn more about island mysteries and know what was happening on barrier islands and particularly Cumberland, I should talk to Carol Ruckdeschel, resident naturalist and wilderness advocate. She has lived on the north end of Cumberland Island permanently since the 1970s and comes by headquarters once a week to pick up mail. Carol had said she could meet my son Sam and I on short notice, and that was too good an opportunity to pass up. We had arranged to meet her at 11 a.m. at Sea Camp dock, on the southwest side of the island where the ferry brings passengers from St. Marys, then talk more the next day at her place.

When I take my family on outings, they accuse me of downplaying the total distance I plan to cover. How far are we hiking? "A few miles," I'll say, vaguely. "Two hours, tops." But then the venture will last three, and my gang will start to turn on me, which is why I always bring chocolate. I think I give a low estimate because I know that if I say we are hiking eight miles and it will take five hours, no one will join me, even though in the end, we will all enjoy the journey. They're on to me, but I do this subconsciously to myself as well. I take advantage of the brain's "optimism bias," which leads us to be more hopeful about future events than might be justified by a careful review of the available information.

And maybe I forgot to check, but I thought the hike from Sea Camp to Brickhill Bluff on Cumberland Island was seven or so miles, which seemed doable, even with packs. Yet, on that summer morning in the air-conditioned room at headquarters for Cumberland Island National Seashore, the ranger informed us that it was really ten. Sam, about to start his first semester of college, with several cross-country running awards behind him, blinked not. I am in decent shape, but the heat index for that late June day was 105. The only one to complain to about the trek would be me.

Hiking companions also like to accuse me of packing the bare minimum, only what we need, no luxuries or excess weight. Sam and I had gathered our gear on the spur of the moment, and I was doing a mental review. Tents and bags? Check. Enough food and water? I thought so. "My Brickhill Bluff people," the ranger said. "Boil your water." I started worrying about the amount of gas in our small backpacker stove. I dug the cylinder of gas out of the pack and shook it. It felt about a quarter full. Enough for cooking, but for boiling water, I wished I had more. Optimism waning.

Carol arrived at noon wearing camo pants and a khaki shirt, pigtails under a sun hat, and rubber boots. Her ATV had been overheating, so she had to stop and let it cool down. She does a weekly beach survey to look for marine animals or other curiosities that have washed ashore, then drops off trash at the ranger center and picks up her mail. I wanted to know if her daily island observations included dune migrations and increased water levels that would point to evidence of climate change. I wondered if a rising sea had affected her beloved sea turtles.

Carol was profiled in "Travels in Georgia," a 1973 *New Yorker* article by John McPhee. In the opening scene of that article, she comforts a dying turtle by the side of the road. After the local sheriff shoots it, she carves up some of the meat and buries some of the eggs in a sandbank, as the turtle herself was about to do; then she takes the rest of the eggs and the meat home to cook. In the decades since, Carol has continued to research sea turtles and fight for the preservation of the island while posting to a website, wildcumberland.org, and her brief brush with fame continued with Will Harlan's 2014 book *Untamed: The Wildest Woman in America and the Fight for Cumberland Island*. Carol was the "wildest woman," and Harlan describes how she shoots a hog that would eat turtle eggs, rolls the turtle over to tag it, and then rides it into the sea: "She straddled the turtle's massive shell and held on to the front edge, riding bareback into the wild waters."

Carol told me she has marked low nests, where the eggs are washed over by the tide, but she does not move them. In general, she does not interfere with natural processes, except when, as in the case of feral hogs released by humans, they interfere with what otherwise would be.

"We've done our damage," she said, as another full ferry of passengers drifted into the dock while we sat on the deck. "There's too many of us," she added, a little wearily. A small jet boat screamed out past the dock. "I'd hate to be a porpoise."

Sitting on the deck at Sea Camp, we talked some more about the health of the island. At that moment, superintendent Gary Ingram, who oversees park policy, was at what she called "wilderness school." She hoped he was learning something, so he could "understand why I'm always a bitch." She was always on his case about wilderness, capital "W," a federal designation, and about keeping Cumberland as wild as possible. About 25 percent of the island is designated as wilderness. A thousand acres are privately held by some twenty families. Recently, one of them proposed to subdivide and develop eighty-seven acres, a proposal Carol would fight.

One issue she and Ingram disagreed about was fire management. She wanted to let the fires burn. The park management wanted to control burns to save particular species such as the longleaf pine, which, Carol said, "doesn't really belong here." She would rather let the scrubs grow. Saw palmetto, staggerbush, and bay all compose a large portion of the north end of the island. After wildfires, the understory is cleared, new light reaches the forest floor, and new growth emerges. Ticks are reduced. Soil fertility is enriched. Birds flourish with the new food and proper cover. Nature finds a way. During a recent lightning storm, she said to herself, "OK, this is it." Remembering this, she let out a mischievous laugh and her brown eyes, like the color of the tidal creeks, grew wide.

I steered the conversation back to climate change. She said she used to be able to walk to Little Cumberland Island, off the north end of the main island, but it was hard to know if that was erosion or sea level rise. Islands are dynamic systems: part of it disappears and forms somewhere else. We agreed to talk more the next day at her place because Sam and I had a long trudge ahead.

On the hike up to Brickhill Bluff, we stopped at Plum Orchard, a twenty-thousand-square-foot mansion built in 1898 by Lucy Carnegie for her fifth son, George. Sam would rather learn about wildlife than capitalist tycoons, but he was grateful for the respite from the heat. So we took the tour, sweaty and grimy from the walk. Sam, shirtless and with his long curly hair tucked into a red bandana, tiptoed on parquet floors through the stately, Gatsbyesque mansion. It is home to an early ice maker, a giant thing in the basement, and an elevator powered by water pressure—both state of the art for 1900. There is a heated, tile swimming pool flanked by squash courts. There are Tiffany lamps designed to look like turtle shells, and red carpet and mahogany tables where formal dinners were the norm, a waiter behind each chair.

The building has a separate stairway and even a separate hallway that servants used. The Carnegies were great believers that the divide between rich and poor was a good and even necessary aspect of social evolution; the concentration of wealth in the hands of the few was, in Andrew Carnegie's words, "essential for the progress of the race." He and other social Darwinists argued on the basis of Darwin's theory of natural selection that the best-adapted humans naturally rose to the top social and economic echelons. It was a rationalization for his own rise to the top tier of his social class, an excuse for exploitation based on a poor understanding of biology.

At the dawn of the twentieth century, robber barons built factories in the cities and mansions on the coasts. Georgia's barrier islands, with their warm weather, quiet seclusion, and abundance of game and cheap land, attracted these millionaires. The Gilded Age came to the Golden Isles. A who's who of corporate tycoons pooled their money to develop Jekyll Island, a couple miles north of Cumberland. Rockefeller, Vanderbilt, Morgan, Pulitzer, and others formed the Jekyll Island Club, perhaps the first gated community, and no uninvited guest could ever step foot on the island. To this day, entrance is fee only. Snubbed by the Jekyll group, the Carnegies purchased Cumberland, the largest and southernmost barrier island.

In the 1960s, developer Charles Fraser wanted to turn the island into another Hilton Head, but the National Park Service offered an even better deal: millions for the land, and the families could continue living there for the rest of their lives. Live there they did, until they succumbed to the same ravages of weather and age as the rest of us. And now we can all enjoy the island, regardless of income. The writer Wallace Stegner once called the National Parks our "best idea." They are also among our most egalitarian.

After filling our water bottles from a spigot on the side of the mansion, we hiked up the road to Brickhill Bluff. We saw armadillos on the way and a mama raccoon with several youngsters. When they spied us, the raccoons all scrambled up different trees, an adaptation to escape predators and one I recognized from my youth: when the bully comes your way, split up.

At the campground, we dumped our gear, stripped off sweaty clothing, and headed for a swim. To get to the water, we had to step over scrambling fiddler crabs and several downed trees. Once imposing and upright, they had toppled over, now grey and ghastly. Whatever high bluff once existed was now an eroding coast, with no root system to protect it. Oyster shells tucked into plastic netting were an attempt at a living shoreline, but they were not holding either, and much of the sand and clay-heavy soil was slipping into Fancy Bluff Creek. A lone egret speared fish near the marsh across the silvery tidal creek while the sun faded over it, casting shadows over the splayed root ball of a fallen live oak.

For dinner, we ate razor clam chowder from our home market, right out of the bag, and used the gas in the cookstove to steam open some clams Sam had found. I didn't want to waste the boiled water so I poured the clam juice into an empty water bottle to cool overnight, just in case we needed it.

Then I lay in a hammock under the live oak canopy and listened to the churr of insects. Despite the sweltering heat, we slept well in our tent to the sounds of cicadas and whip-poor-wills. In the night, I went out to investigate a sawing sound, which turned out to be an armadillo scratching against a palm. We woke with red spots on our arms, from no-see-ums, ghosts of the night. By morning, I had found so many ticks on my body I started trying to scrape off freckles.

After breakfast, we hiked a few more miles to the north end of the island. Carol lives near the First African Baptist Church, a one-room structure with white clapboard siding and tin roof, where John F. Kennedy Jr. married Carolyn Bessette in September 1996 to avoid press and photographers. From the church, you can see Carol's cabin, weathered siding with pieces of driftwood attached to it. The yard is scattered with rusted tubs, farm tools, crates, broken-down carts, and buckets stinking of sea life. Maceration is Carol's preferred method to skeletonize marine animals. After soaking the carcasses, she bleaches the bones in the sun.

Carol joined us at her picnic table beside her garden and chickens, telling us how she arrived on Cumberland. She had been making visits since the 1960s but moved there full time a few months after the National Park Service purchased much of the island in 1972. She worked for the Candlers, heirs to the Coca-Cola fortune, and what was the Cumberland Island Hotel, a resort built in the 1870s. At one time, there were some four hundred black families on the island, some of whom worked at the hotel. They lived on the north end, on the least desirable land. There were still some collapsing shanties in "the settlement" when Carol arrived in the 1970s.

Days, she polished silver. At night, she waited for sea turtles. Over time, she would become a leading expert on sea turtle mortality and conservation, publishing scientific articles and a book, *Sea Turtles of the Atlantic and Gulf Coasts of the United States* (2006).

She described how sea turtles go back to their natal beach—the place they were born. I wondered if rising seas would affect their habitat, but Carol told me they are "programmed to be flexible." Loggerhead sea turtles live for thirty-five years before they sexually mature, and some of their current success is the result of conservationists' efforts over thirty to forty years ago to protect their breeding sites. In the summer of 2016, the park found eight hundred nests on Cumberland. Another naturalist, Stacia Hendricks of Little St. Simons Island (another barrier island), told me they found eighty-one nests in the middle of June alone—a high tally for early in the season. Hendricks seemed to share Carol's view of turtles, that they are built for survival. Turtles have an antifreeze in their blood that allows them to tolerate cold, and some sea turtles can collapse their lungs, an adaptation that helps them avoid the bends when diving to the bottom of the sea floor. Sea turtles, Hendricks argued, have lived for millions of years and will continue to do so with a kind of slow, steadfast stamina.

Some researchers are not so hopeful about the sea turtles' prospects for long-term survival, however. For animals that use natal homing, if that home disappears, will they know what to do? Some will "stray" and find a new home, but mostly they are squeezed. And because nest temperature affects sex selection, climate change could also affect turtles' incubation. As a group of interns from the Georgia Sea Turtle Center on Jekyll Island told me, "the ladies like it hot." Hotter sands caused by a warming climate and increasing drought could cause greater numbers of sea turtles to be born female, which is not at all good for the long-term viability of the species.

After we talked, Carol gave us a tour of her "museum." Though the outside was a clutter of objects, the twenty by thirty plywood, freestanding building was clean and orderly inside, with bright walls but no windows. And it was air conditioned. She and her longtime research partner, the late Robert Shoop, who retired from academic herpetology to live on the island, created it. It's not open for tourists but was built to house her life's work of scouring the beaches and nearby forests for the dead. "I always tell people, 'find anything dead, bring it to me,'" Carol said, cheerfully. Specimens were neatly arranged, labeled, indexed on handmade shelves. A card catalog housed insects. On the wall, hung skulls and skeletons of every creature on Cumberland Island, identified by both scientific nomenclature and common name. Over the entrance perched the perforated baleen of a juvenile humpback whale. In the back, skeletons of sea turtles were stacked like nesting boxes.

Carol called out stuff she wanted us to see. A coyote puppy. The biggest hawksbill turtle ever recorded in the state. A painted bunting. A screech owl. Sam enjoyed the snakes coiled tight in jars, milky-eyed. A small egret sat in an olive jar. Touring through here, pulling out drawers and trays filled with bodies, is like walking through a catacomb, and Carol is a kind of caretaker of the dead.

Some of her collection has been sent to the Smithsonian, a fact she delights in telling. But much of it, all labeled and arranged, the stomach contents analyzed, may have no purpose other than scientific curiosity.

Scientists journey from far and wide to sort through this treasure. Perhaps the mystery of how climate change has affected sea turtle diet is in here somewhere, the indexed stomach contents, data about diet, and dates. Hours and hours of labor are packed into these bags of bones, puzzle pieces for a picture yet to be made clear.

"We hoped to make their death useful," Carol said, turning out the lights. "You never know what may turn out to be valuable."

While I signed the guest book, we talked some about TEDs—turtle excluder devices for shrimp trawlers (hatch doors at the bottom that allow turtles to swim free from shrimp nets)—which Carol had a hand in implementing and which have been required on large trawlers since 1987. From her perspective, they did not reduce strandings, when turtles are harmed by nets and end up injured (or dead) on the beach, possibly because shrimpers, who resisted the regulation, skirted the law. Carol was as busy as ever picking up carcasses in the early 1990s. What has helped is an economic downturn in the shrimping industry, brought on by adaptations in aquaculture; increasingly, shrimp are grown in ponds without a need for boats or fuel. Nest counts still go up and down, mysteriously, but the overall trend in Georgia is up by as much as 3 percent annually over the past two decades, according to Mark Dodd of the Georgia Department of Natural Resources. There were some 3,200 loggerhead nests on Georgia beaches in 2016, up from a low of 470 in the early 1990s. This trend is holding in other southeastern states too, though the turtles are still listed as threatened in this region.

The trail to Stafford Beach, seven miles to the south, passes through a saw palmetto and live oak forest, dark and eerily gothic. Despite the positive news about sea turtle nesting trends, a pensive mood overtook me. Trudging through leaf litter and pine needles, I thought about "useful death," Carol's notion that something good could come from the creatures she studied. Maybe the dead trees last night or the tour through the catacomb in the morning influenced my thoughts, seeming to rise and fall with the trail. So much dead and dying. So much simultaneously alive and buzzing around me.

We walked through a wetland where there were quite a few tall dead trees, gray and weathered, ghostlike. Scientists call these "ghost forests." Those trees on the Brickhill Bluff coast could have toppled into the water because of boat wakes, but they could also be the result of storms or rising seas, which degrade the peat and soil, destroying root systems.

Chester "CJ" Jackson, a coastal geologist at Georgia Southern University, has studied shorelines and estuaries of the Georgia islands, using a software called AMBUR, Analyzing Moving Boundaries Using "R" (where R is a software for statistical computing), developed through funding from Sea Grant, a NOAA program dedicated to coastal research. He has been digitizing the coast, comparing the modern coastline with historical data. Using aerial photographs and maps and plotting the change, Jackson determined rates of erosion and accretion along the coast. Red lines showed erosion on islands; blue, accretion. The southern end of the island tended toward blue, while the northern tip was red. But he also found erosion on the backside of the island and up into the marshes. At Brickhill Bluff, he plotted an average rate of erosion of a half meter per year over a twenty-year period.

Of some seventy-seven thousand transect points Jackson plotted in Georgia, over half experienced erosion of up to one-half meter per year. But whereas coastal points experienced some net gain, deposition in the marshes was not keeping up with erosion.

I asked Gregory Noe, a research ecologist who has been studying the impacts of dying cypress along the Savannah River, about the dying trees in these coastal marshes. "Those trees are a dramatic expression of climate change," he said. The river brings sediment to the marsh, combining with dead plants and peat, a natural accretion, but the swamps are losing the race with the rising sea. Drought is also a factor, for if the rivers do not carry enough fresh water, the trees are affected. Noe told me these "ghost forest swamps are harbingers of what is to come." Sea level rise will make things worse. On Cumberland, according to a 2016 study prepared by the U.S Geological Survey, the average retreat would be fifteen meters by 2050 and approximately thirty-seven meters by 2100. That's bad news for the islands but also for coastal communities for which the islands provide a crucial buffer against storm surge.

I kept thinking about a "useful death" and wondered if these dead trees, and not sea turtles, were the necessary visual to communicate the danger of a rising sea. Carol herself studied death to better understand life. I asked her, months later by email, if she stayed optimistic despite her daily contact with death, the unrelenting march of it. "I am not optimistic about our species," she replied, adding a frown emoji.

By late afternoon, I had again fallen under the trance of the live oak understory, pine needles and sand crunching underfoot, Spanish moss brushing my shoulder. I had some idea where we were, but my map was sweat soaked in my back pocket. The day was even hotter than the one before, and we had glugged most of our water. I was down to clam juice—salty, a little fishy, not altogether bad—which would, I hoped, provide necessary electrolytes. I thought of the water in the marshes, the chicken soup–like broth that feeds shellfish and birds.

Sam asked how much farther we had to go. I was ready to be at camp too, so sweaty a rash developed between my legs. If we ran out of fuel, the heat of my inner thighs might stand in. In these swampy parts of the island, the mosquitoes multiplied. I scratched around the tops of my socks for some invisible bug, possibly chiggers. Paradise was not supposed to itch this much.

That nature is a static, unchanging paradise is an idea created by culture. Where I walked, the land looked like wilderness, but in fact, former inhabitants had trodden and cleared it. Cattle and hogs were let loose, trampling the ground underfoot. Horses still roam, grazing, a sore spot with Carol, who would have them eradicated because they do not belong there. Near the beach where we were headed, at Stafford Plantation, there were ruins of slave quarters, lone chimneys from plantation days. The constant on this island is a buzzing, humming change.

Out of a pastoral impulse grew our national parks; our guide on the ferry ride to Cumberland had reminded us of the history. In 1916, when Congress created the National Park Service, it charged the agency to leave the parks "unimpaired for the enjoyment of future generations," though what *unimpaired* meant was never defined. In the 1960s, Secretary of the Interior Stewart Udall, who oversaw the purchase of Cumberland, had wildlife biologist A. Starker Leopold, son of conservationist Aldo Leopold, chair a study. The resulting report directed the National Park Service to maintain the original "biotic associations" that existed before European settlement, each park a "vignette of primitive America." What the landscape was before development or destruction.

We dumped our packs at camp, quickly pitched tents, filled empty water bottles, and then strode over the dunes to the shore. No condos stole the view, no boardwalk or Ferris wheel, and very few people. Instead, just dunes and the rolling surf. We swam in the cool ocean, but the insides of my thighs burned in the salty water, no cool relief to counter the sting. We came to meet Carol and learn about sea turtles but also for this, a spectacular view of the wild ocean coast. It might indeed be Leopold's vignette of what things once were.

In 2011, Park Service director Jonathan Jarvis had a panel reexamine the Leopold report. "Revisiting Leopold" proposed a new set of goals. Rather than re-create primitive vignettes, the National Park Service would manage for "continuous change that is not yet fully understood." The report does not reflect official policy, but it acknowledges that changes are underway and that the parks' infrastructure will have to adapt to those changes. Here on the seashore, that means the park will not try to fight the inevitable but may move as the island moves. It means understanding vulnerabilities so a plan can be put in place. It is a recognition that not all can be saved. "We can't control all that lies ahead, but we will prepare for it," Superintendent Gary Ingram told me. "Being a dynamic island, we're along for the ride."

Back at camp, we used the last of our gas to heat water for our dried rice, cheese, and broccoli mixture. Snacking on protein bars and nuts, we eyed the stove, willing it not to peter out before heating just enough water to pour into the mixture. We were hungry enough to eat the packaging.

Staring in anticipation of the boiling water, I reflected on first learning of climate change in college thirty years ago, growing weary of how little we had done. Sam would soon begin college, a future in front of him that will require serious thinking about climate change. We talked about optimism bias and how it was built into us. Was our Pollyanna-ish belief that things will turn out fine responsible for the reckless building we did on the coast? Was it what kept us from tackling climate change head on? Our brains tilt toward the positive, but they mislead us. They tell us stories—the ones we want to hear—that help us cope with the unpredictable. Optimism bias is a problem with our risk calculus: "It won't happen to me." Humans seem better equipped to imagine a bright future than confront a likely one; we are comforted more by illusions than hard data.

During most of our evolution, this adaptive trait made sense. Bad weather, accidents were beyond our control. Tomorrow would be different. Such an outlook was good for mental and even physical health, as it reduced stress. But for the first time in history, we can predict the decades-long consequences of our past activities. Analysts name optimism bias as one of the core causes of the financial collapse of 2008. Many hope we will figure our way out of climate change before too much damage is done. According to Yale University's Program on Climate Change Communication, while 70 percent of Americans say global warming is happening, only 40 percent think it will harm them personally.

When we give ourselves over to optimism, we give up our agency and put faith in something else—technology, a supreme being, aliens from outer space, magical thinking. Rejecting optimism, we stop being hopeful that the awful situation we are in will somehow resolve itself, and we start working to change it. After blind optimism comes thoughtful action. Without that optimism trait, the human species might still be huddled in some cave. But it's we who have to do the work. We who have to double-check that there is enough fuel for the fire.

Before we tucked ourselves into our bags, Sam set an alarm for midnight; we were planning to walk the beach and look for nesting sea turtles. He woke me from a deep sleep, and we were both too groggy to find the red filter, buried somewhere in the pack, for the flashlight. (Lights can deter the turtles from coming ashore.) Bleary with sleep, we walked in darkness, me like a bowlegged rodeo cowboy.

We traipsed along the water's edge, in the soft sand above the tide line, until we spied a dark form before us—perhaps a turtle burrowing? But no, it was two people locked in an embrace. We kept walking, watching the shore, listening to the waves, also scanning the dunes in case one was up there. Our hopes were dimming for an encounter, and I realized we probably didn't have a good idea where we had come out. Perhaps we would find our way back to our tent by memory, not unlike a sea turtle.

Then we saw one. Waves splashed against it, kicking up spray as it crawled on flippers, making treads in the sand. Heave and ho, one appendage tilting the massive, ancient body—some three hundred pounds—up, forward, then down, and repeating on the other side, like a cart with oblong wheels. When we encroached, for a minute its progress toward the dune halted. The turtle seemed to turn some with the line of the tide, as if being chased. So we moved out of its way, even tried to look away, as if that would help.

We had the impulse to move closer, to get right down to the barnacles and seaweed stuck to its frame. Maybe even to touch it. But we also wanted to keep away, to clear the path, to sit on the sidelines, and cheer for its success. Its journey had been too long for it to turn back now. "Big mama," one of us whisper-shouted, the other in silent awe. A thing as big as a bathtub emerged from the inky depths like a shadow of the distant world.

I hadn't wanted to wake up, but I was glad Sam had set the alarm. It was an awesome sight, a primeval creature heaving itself to shore, aided by ancient, natural processes but also by protective measures begun by humans, including Carol, some forty years in the making. That's what we were witnessing, something both ancient and relatively new. It's the same with sea level rise.

Recognizing its advances will require being attentive, alert, even when we don't want to be. Meanwhile, the humble turtle plods on, sticking its neck out, enduring drought, volcanoes, glaciers, and a meteor strike. It has survived mass extinctions that wiped out much faster-moving species. Slow and steady won the evolutionary race. But changes are happening now more quickly than most species can adapt to.

I was buoyed by the sight of it, a symbol of the resilience of nature. To endure, like the turtle, we ourselves will have to adapt. An alarm is ringing. A nearly invisible thing moves toward us in the night. Do not be misled about the distance we have yet to go.

6

The Octopus in the Basement

Surreal Matters in the Sunshine State, Florida

Much about Florida hardly seems real. At least that was my impression when we arrived in the middle of December, temperatures in the mideighties. There were outward signs of holidays—inflatable snowmen and reindeer and lighted palms instead of spruce trees—but weather-wise Florida seemed not to be participating in winter.

I had turned in my grades and my son finished his exams, so perhaps we were decompressing after busy semesters, but suddenly, we found ourselves in a kind of tropical paradise, sand and sun. It did not seem real.

Before heading further south, we visited a colleague and friend who had recently retired. He built his dream house several blocks from the beach and several in from Mosquito Lagoon, just north of Cape Canaveral National Seashore. We were a college student, a midcareer professional, and a retiree, all seeing the world somewhat differently. Before he retired, Don could be known to complain about college students, their lack of preparation for writing or editing, but he was feeling stress free now. What to do tomorrow? Walk the beach? Cast a line? In some ways, his life was similar to a college student's—meeting new friends, taking up hobbies—only he could skip the classes.

As we drove through his neighborhood, we noticed that either Santa and his inflatable reindeer could be found on rooftops or a team of contractors roped in were up there, repairing the damage from Hurricane Matthew. Our friend suffered only minor damage to a soffit and some to the new floor as water blew in under the second-floor door.

Sam and I traveled to Florida once before. My daughter devoured the Harry Potter books and was anxious to visit Universal Studios and their re-created world of the fictional one, a replica of a replica. Instead of touring the park, Sam and I canoed through Rock Springs, a clear river that twists and turns through lush vegetation, pristine. We saw strange fish and birds we could not name. Around one corner, a deer bedded down in the grass at water's edge. We scared it off, but it left an imprint. My son remembered it long after and wanted to return to Florida in part to return to a landscape that made such a similar impression. But reality can be a hard match for imagination.

Many of the flowering plants added to the show. We saw no orange blossoms but many others on aromatic display. Looking for the fountain of youth, Ponce de Leon found not youthful springs but spring flowers, naming the state after the florid vegetation. Some of these plants we witnessed were invasive, like Brazilian pepper, native to South America but brought into Florida in the nineteenth century as an ornamental. It likes mangrove forests; mammals and birds spread it. Brazilian pepper berries produce a narcotic or toxic effect on native birds and wildlife during some parts of the year.

Florida has made national news for invasives. Among the most wanted are Giant African snail, Cuban tree frog, green iguana, and Burmese python, often introduced by pet owners. People buy the snakes as babies but do not realize how fast and large they grow—up to twenty feet. When owners run out of space, or a supply of rabbits, they simply open the door, a terrible strategy in a region where the snakes have no natural predators. The problem with exotics is especially pronounced near the Everglades, with a tropical climate and the proximity to a major airport. By one 2004 estimate, 26 percent of all resident mammals, birds, reptiles, amphibians, and fishes were not native to South Florida. Steve Traxler, of the U.S. Fish and Wildlife Service, told me the exotic pet trade is as profitable as the drug trade. And exotics, like drugs, do billions in damage.

And yet, Florida is rich in biodiversity. A peninsula buffered by the ocean, it stayed warm during the Ice Age more than ten thousand years ago. For that reason, it has a high number of endemic species.

Of course, another kind of species "invades" Florida daily. After we crossed the state line, we heard on local radio that Florida was second in the nation in population growth (behind Utah). The state regularly leads the nation in in-migration, adding nearly one thousand people daily. Pythons have not stopped the flow of people.

The night we arrived, we sat outside at picnic tables, eating fish tacos and sipping beer. For dessert, ice cream cones with sprinkles, a summertime ritual for us.

Adding to the sense of an alternate reality—it was the first trip I had taken since the election. Donald Trump launched his candidacy from an escalator in Trump Tower. Eighteen months later, he swept away many preconceived notions of what it took to win the White House, and left many writhing on the floor.

Weeks later, his transition was beginning to resemble a kind of reality TV show, where contestants for positions in his cabinet passed through the doors and up the elevator, an invisible ascent before they met their destiny.

Al Gore met with Trump for an hour and a half, describing to reporters afterward an "extremely interesting conversation," giving some hope that Trump had an "open mind" on climate change, as he told reporters. Two days later, after the Gore head fake, Trump tapped a climate change denier who worked as a hired gun for fossil fuel companies to head up the Environmental Protection Agency. And a week later, Trump picked the head of ExxonMobil as secretary of state. As CEO of ExxonMobil, Rex Tillerson viewed climate change as a threat to the company's bottom line rather than a signal to change direction. Most other fossil fuel companies began to adapt, investing in solar and wind, recognizing a changing business environment.

Mr. Trump, who has criticized the established science of human-caused global warming as a hoax, canceled U.S. participation in the Paris accord, which committed nearly every nation to take action to fight climate change, and attacked Mr. Obama's signature global warming policy, the Clean Power Plan, as a "war on coal." Scott Pruitt, director of the Environmental Protection Agency, has been in lock step with those views. "Scientists continue to disagree about the degree and extent of global warming and its connection to the actions of mankind," he wrote in the *National Review*. Only they do not disagree. President Trump later said "nobody knows" what is going on with climate change, not understanding scientific uncertainty, which science deals with. It weighs the evidence and decides which of several possible answers has the most support. Where uncertainties remain, scientists know where to investigate next. Regarding climate change, there is no uncertainty that it is happening and that it is accelerated by the burning of fossil fuels. The only uncertainty is what will happen and when and how people and communities (and their governments) should respond.

Trump later added that other countries are "eating our lunch." What is certain, in his mind, is that climate change policies hamper business competitiveness. "Nobody really knows" is an argument of convenience, as doing something about climate change goes against his economic views. In fact, we do not really know if addressing climate change will hurt or help the economy. Investing in renewables and adaption strategies may very well stave off disaster and fiscal collapse. Trump has it backward. Other countries, like China, will "eat our lunch" when it comes renewable energy sources, adapting to a new economy.

In the morning, Don took us sightseeing to Turtle Mound, the largest oyster midden on the mainland coast. Standing at over fifty feet, it was a refuse pile of oyster shells for the Timucuan people, over a thousand years ago. These people lived inseparably connected with the estuary, reflecting a lifestyle tied not to agriculture but to hunting and gathering.

I asked the rangers in the visiting center if the mound could have had other purposes, such as shelter from a storm, and the naturalist told me likely no. The park was in the process of a restoration project as part of the trail on the lagoon side experienced erosion. They planted red mangrove and added oyster bags. Oysters to protect the prehistoric mound of oysters.

Outside, the flag flew at half-mast in honor of John Glenn, the first American to orbit the Earth. In his tribute, President Obama wrote that when Glenn "blasted off from Cape Canaveral atop an Atlas rocket in 1962, he lifted the hopes of a nation."

We drove on down the barrier island that is Cape Canaveral National Seashore to the southern tip. We parked and walked the beach, though our friend told us that we would not go far. If we kept walking, we entered the unofficial clothing optional beach. A sign said "nudity is not allowed," but it was a federal area and county laws were not strictly enforced. Rangers are said to "look the other way" unless they detect lascivious behavior. We didn't walk far because nothing would be weirder than to stumble into a group of adults without clothes. Then again, perhaps that is how nature would want us. *Naturists* is the preferred term for that particular lifestyle, while nudism is simply the act of being naked.

It was odd to consider, the naturists in primitive buff side by side the most expensive, technically complex enterprise ever. But what attracted the naturists to the beach was the same thing that attracted the space program years ago: a stretch of undeveloped beach facing east over the ocean and with warm, equatorial weather.

After Russia fired off the Sputnik, the cape came to be the place where NASA built launchpads for catching up to the Soviets. NASA has two active launchpads, 39A and 39B, used for the shuttle and in the future for the Space Launch System and SpaceX's heavy lift vehicle. The coastal area between 39A and B has experienced increased erosion due to climate change and sea level rise as waves push water high up on the beach. High tides have crept closer to a service road under which lay some of the center's natural gas, communications, liquefied rocket fuel, and water lines, just four feet above sea level.

What to do about coastal erosion? Throughout most of Florida, the typical answer is dredging sand from underwater and using it to rebuild the shoreline, called beach renourishment. According to journalist Craig Pittman of the *Tampa Bay Times*, thirty-five of Florida's sixty-seven counties have used taxpayer money to artificially enhance their beaches in this way, "plumping them up like a fading star injecting collagen in her too-thin lips."

NASA decided to do things differently. Instead of building back the shoreline, the agency used beach sand from a project at nearby Patrick Air Force Base to build a second, mile-long line of dunes inland from the area where the erosion was occurring.

"Renourishment would be much more expensive," Nancy Bray, director of Spaceport Integration and Services for Kennedy Space Center told Pittman. Besides, she pointed out, the rising sea would just wipe out the built-up beach all over again. Even better, the second dune system created habitat for several endangered and threatened species that call Cape Canaveral home.

Still, Bray and colleagues at Kennedy are working on a "managed retreat." They build up the shore dunes but also reserve land farther inland for additional launch sites, assuming demand (and finances) materialize. Water is not lapping at launchpads yet, but sea level rise does affect the majority of NASA centers that are on the coast.

During the election, two advisors to the Trump campaign, Peter Navarro, an economist at the University of California, Irvine, and Robert Walker, a former member of Congress from southeastern Pennsylvania who was chairman of the House Science Committee from 1995 to 1997 and chairman of the Commission on the Future of the U.S. Aerospace Industry, posted an article for Spacenews.com that said: "NASA should be focused primarily on deep space activities rather than Earth-centric work that is better handled by other agencies."

The threat of climate change was thrust into the public consciousness in June 1988, when NASA scientist James Hansen told a congressional committee that researchers were 99 percent certain that humans were warming the planet. "The greenhouse effect has been detected, and it is changing our climate now," he said in his testimony. Since then, NASA has played a leading role in researching climate change and educating the public about it. The space agency's satellites track melting ice sheets and rising seas, and its scientists crunch the data showing how quickly the Earth is warming. NASA studies not only "deep space" planets but the planet Earth. NASA monitors temperature and weather (including hurricanes), shows how changes to atmosphere trap more heat, and displays through infrared imagery the impacts of climate change to ecosystems and communities. The National Aeronautics and Space Act of 1958 provided "research into problems of flight within and outside the earth's atmosphere, and for other purposes." Its goal was the "expansion of human knowledge of phenomena in the atmosphere and space."

One of the many advances the space program gave us is the smoke detector. Writer and climate activist Bill McKibben has said that defunding earth science efforts of NASA, which monitor carbon dioxide levels and monitor ice sheets, could be compared to tearing out the smoke detector just as the house is about to burn.

Two politicians from Florida are among the many who have used the phrase "I am not a scientist" when asked about climate change: Sen. Marco Rubio and Gov. Rick Scott. Neither are they economists, but that has not held them back from opining on tax policy. Scott agreed to meet with several scientists—Ben Kirtman, professor of atmospheric science at the University of Miami; David Hastings, professor of marine science and chemistry at Eckerd College; and Jeff Chanton, professor of oceanography at Florida State University—so they could explain the issue. But not much seemed to come of the meeting. Hastings said the governor did not ask the scientists "substantial questions" or respond to the presentation in a way that showed "substantial or significant concern on his part." Chanton told me that Scott "listened perfunctorily," if he listened it all, but spent a good portion of the thirty-minute meeting asking about their students and the courses they teach. Scott did not take questions after the meeting but several days later offered these comments: "I'm a business guy. I'm a solutions person. So my focus is: we know there's issues out there—sea level rise—so let's focus on how we solve it."

Science can help solve this and many problems. Chanton has pointed out that as we are just learning of the effects of increased carbon dioxide and other gases in our atmosphere, 150 years ago we did not understand the link between microbial bacteria and disease. An understanding of science benefits our well-being and allows us to live more comfortable, productive lives. Speaking about that spirit of ingenuity, President Obama said in his farewell speech, that it "made us an economic powerhouse—the spirit that took flight at Kitty Hawk and Cape Canaveral; the spirit that that cures disease and put a computer in every pocket." Perhaps it helps us understand Zika, a virus that has affected Floridians and that may be bolstered by climate change.

For the most part, it was difficult for me, in balmy Florida, to lead a productive life. Comfortable yes. In the afternoon, we swam in the ocean. Then we lounged around the house beside the bougainvillea and lemon tree. The lush fantasia that is Florida pulls you in. A narcotic effect of some exotic.

But I managed to pull myself away from the lawn chair and get to a meeting with Jason Evans. He has worked on a number of adaptation plans for communities in the Southeast, including one for Tybee Island, Georgia; Hyde County, North Carolina; and Monroe County and the Village of Islamorada, on the Florida Keys. He has lived in Florida most of his adult life. Over coffee, we talked some about the recent history of environmental policy and climate change in Florida.

The thumbnail sketch I could piece together went something like this. Under Jeb Bush (1999–2007), there was an emphasis on water quality, the health of the springs, but not necessarily climate change. Under Charlie Crisp (2001–11), there was pro–Everglades restoration and more on climate change. Then the wheels started to come off. Governor Scott began to dismantle the framework that had been set up. People in the water management districts who worked on climate change were let go, demoted, or muzzled. Under Scott, the Department of Community Affairs, which used to put some brakes on runaway growth, was swallowed up by the Department of Economic Opportunity.

In 2015, the Florida Center for Investigative Reporting broke a story about how the state's Department of Environmental Protection (DEP) had an unofficial policy to not use the terms *climate change* or *global warming* in communications, emails, or reports. With more than one thousand miles of coastline and some 2.5 million people living within four feet of the high tide line, low-lying Florida is among the places in the world most vulnerable to climate change and the resulting sea level rise. The irony seemed delicious. But the DEP's press secretary denied such a policy, and a spokesperson in the Scott administration denied that it came from a directive above (the governor himself denied it). However, the DEP has no program for climate change either. Reed Noss, a retired professor of conservation biology at the University of Central Florida, told me people he knows who work at Florida Sea Grant "cannot talk about climate change" when talking to the public. However, sometimes this is a matter of knowing your audience. Evans uses tactics of risk communication when working with communities to develop plans, which include knowing his audience and their values, establishing trust and credibility with them, and offering the options without scaring them to death.

While Evans's son Isaac pulled the chocolate chips out of a muffin, we talked some more about planning. Isaac is six, but Florida's largest voting bloc skews to the other end of the age spectrum. The most reliable voting bloc in Florida are senior citizens. That is one reason why Florida has trouble passing bonds for mass transit. The ones who pay for it may not be around to enjoy it. Jim Beever, another planner who works with communities on adaptation plans in the Naples/Fort Myers area, told me he sees a lot of retirees in his public meetings. They have the time to give to the process, whereas people who are working or raising families have other commitments. Planning for Beever is a matter of public participation. He wants to help people get to particular goals, but they tell him what they want. He does not try to force a concept. Still, he gives the data. "I take my time. I tell people there's a natural component and an accelerated one, that humans have added to climate change in significant ways." "Belief in climate change is optional," he likes to tell communities, "but participation is mandatory."

The other voting bloc with the power to do something is the money along the coast. Evans told me that when he does adaptation plans, he uses a tool provided by NOAA to do cost-benefit calculations. He plugs in property values and comes up with a better understanding of a damage assessment. This can mean that resources are steered toward those dollars. Should they be? And should all of us pay for those who build on the coast? Fixing all the infrastructure required to fortify beach communities, such as adding storm water pumps, could be a "recipe for fiscal disaster." Much of the major financial investment in Florida is below ten feet of elevation. According to the Organization for Economic Cooperation and Development, Miami is the most exposed city in the world in terms of property damage with some $416 billion in assets at risk.

Some communities have other priorities, like homelessness, and cannot afford adaptation planning. Some, like Pelican Cove, a private development, pooled their own money to pay Jim Beever for an adaptation plan. Beever gave them one with "no regret solutions," meaning the changes are beneficial even in the absence of climate change and where the costs of adaptation are relatively low when compared to the results of the adaptations.

We left the coffee shop and wandered over to the Marine Discovery Center on the Indian River. They restore oyster reefs and mangroves, eradicate invasive species, monitor water quality, and clean up the coast. Red mangrove trees have specialized adaptations to survive in the extreme conditions of estuaries. Red mangroves have reddish prop roots descending from the trunk and branches, propping them up during waves and tides. They also have aerial roots, sticking out of the water and taking up oxygen. Mangrove roots can also filter out the salt, allowing them to live in harsh saline environments. Black and white mangroves excrete salt from their leaves. The red mangrove is the most salt tolerant, found closest to the sea. Further inland at a slightly higher elevation, black mangroves take over, rarely flooded by high tide. Finally, white mangrove and buttonwood face inland and are the least salt tolerant.

One indicator of climate change is that the mangrove forests are moving north. We had seen some mangrove seedlings while walking the beach earlier in the day. Cold snaps of twenty-five degrees or less stop mangroves, but with few hard freezes, mangroves have been free to expand into new territories like the coastline near St. Augustine. Since they buffer against storms and provide vital habitat for aquatic species, perhaps this is a benefit of climate change. Certainly, they store carbon, up to *four times* more carbon per hectare than other tropical forests according to a 2011 study. The last deep freeze Evans remembered was 1989, Christmas Day.

Evans is lean, likely a runner, and happiest in the field. His son played around near an area designed to show different approaches to erosion control: permeable pavers, grass, and a tiered wall. From a small perch, we took in the Indian River, also affected by climate change. In July, Mosquito Lagoon experienced an algal bloom, the color of a "nuclear Yoo-hoo" said one witness. Climate change contributes to the proliferation of these blooms and fish die-offs because warmer water triggers them and holds less dissolved oxygen. When murky green, it's not the color of water we would want kids to play in. "It's depressing work," he said. "Stuff will live," he continued, as if struggling to find the silver lining in an era of climate change, but you "can't fortify yourself against the ocean permanently." We talked some more about those who are "not scientists" but who have children. Rubio has four. Scott has grandchildren. "It's going to be interesting to watch," Evans said, looking at Isaac tracing a stick in the sand. "Perhaps natural systems are not as sensitive in the near term as we think," he added, "but I'm not betting on it."

After our meeting with Evans, we drove to Lake Woodruff National Wildlife Refuge near DeLeon Springs. Evans talked with Sam about his studies in wildlife conservation and told us about the place. Sam was interested in joining me on this trip in part because Florida is such a welcoming environment for cold-blooded reptiles. Like the state's retirees, snakes like their environs warm. And Woodruff was said to be home to pygmy rattlesnakes or Florida ground rattlers. Pygmy rattlesnakes are small by rattlesnake comparison, but they are quick to strike and feisty, yet their bite is rarely fatal. Still, while Sam poked around under logs and palmetto leaves, I looked up the location of the nearest hospital. Pygmies love the hardwood hammocks along the borders of freshwater marshes and cypress swamps. We looked for several hours, hearing leaves moving and faint buzzes we thought could be rattles, but he found nothing but a velvety skin camouflaged to blend with the surroundings. When the snakes are found, they are often in a coil, like a small drink coaster. Rather than hunt, pygmies often wait for prey to the come to them. And they use their small rattle in a distinct way: not to ward off danger but lure in food, like small lizards, that would be attracted to the sound of it, an insect-like buzz.

It was a hot, breathless, sweaty day, and I imagined the namesake of DeLeon Springs walking around in his armor, which would have offered protection from the arrows of natives and maybe rattlesnake bikes—but not mosquitoes—and certainly no relief from heat. After an hour or so of Sam looking, I got into the act, scanning the ground for a slender form. I was glad to share in the finding, though I wanted nothing to do with the handling. Looking for one reminded me of searching the ground for morel mushrooms. You have this vision of the thing you want to find, and intently sweep the forest floor until the real thing comes into view. I felt I was using something like Terminator vision, screen zeroing in on the prototype, or maybe snake eyes, trying to pick up hues and shades, undetectable movement. Pygmy rattlesnakes are camouflaged, so I kept conjuring up the slender form against the tan fronds and dark soil, seeing serpents where there were none.

Though we found no rattlesnakes, we did find a ribbon snake when we least expected it while we ate lunch on a small bridge. We spied numerous heron, some coots, whose black bodies we mistook for alligators, and an alligator that was an alligator, lying low in the grasses at the canal edge. We also came across a first for us, a gopher tortoise, a threatened species. They love sandy soils and have large feet for digging burrows. These burrows help them survive during brushfires, and they are a keystone species because so many other species use these burrows for habitat.

While searching my phone for nearby hospitals, I stumbled into a news article: "A missing man in DeLeon Springs was found dead, jammed head-first in a gopher tortoise hole on land where he kept caged alligators."

Driving in Florida remained an adventure. I half expected little old ladies peering over steering wheels, but people in Florida drive fast. And as you approach South Florida, the traffic increases exponentially. Expensive-brand black cars passed us on both the right and left, dodging other traffic, in a tremendous hurry to somewhere—very unlike a tortoise. Our rental car had been upgraded from compact to intermediate. Now styling in a black Buick, we almost fit in.

While we walked around Turtle Mound the previous day, this day we drove past the highest point in South Florida, a garbage mound rising to some 225 feet known locally as Mount Trashmore. After receiving complaints about odors, and after being denied a permit to expand its height to 280 feet, Waste Management renamed the site from catchy "North Broward County Resource Recovery and Central Disposal Sanitary Landfill" to the more concise "Monarch Hill Renewable Energy Park." At the unveiling of the name change, and to reflect its "commitment to environmental sustainability," so says the company's website, they had local students release three hundred monarch butterflies "to find new homes among the 200 butterfly attracting vines placed along the park's perimeter." As anyone who composts knows, butterflies dig trash.

Where was all that trash and traffic coming from? In the twentieth century, Florida experienced a population surge unparalleled in the United States. During the 1950s and 1960s, Florida grew four times as fast as the rest of the nation. What made the rapid urban expansion possible was the draining of portions of the Everglades. From 1947 to 1971, the Central and South Florida Flood Control Project authorized construction of some 1,400 miles of canals and flood control structures, a massive plumbing system of levees, dams and ditches from south of Orlando to Florida Bay. They drained the swamp. Today, the Everglades receive less than one-third of its historic flow, and what water it does receive is impaired by fertilizer and other runoff.

That night, looking for a place to eat, we ended up in Hollywood, Florida, driving down Hollywood Boulevard, not the one in California but its twin, the dream of a developer who wanted to create a matching movie paradise but on the East Coast. We stayed on Route 1—stretching from Key West to Fort Kent, Maine—among remnants of early car culture such as motor courts, though now with corporate franchises.

Broward County gets its name from one Napoleon Bonaparte Broward who won the 1905 governor's race on a platform of draining the swamp, literally. A major flood occurred in 1903, destroying the majority of the crops and farms in the Everglades watershed. It became such an important issue that the governor's election of 1905 was perched atop a platform of flood control. The elected Napoleon Bonaparte Broward won primarily on a promise to "drain that abominable, pestilence-ridden swamp." He called the Everglades "that fabulous muck." So began the long series of drainage projects to reclaim land for agriculture and development.

In the morning, I drove over to a seawall I wanted to see that is notorious for being low. Hollywood Boulevard splits two manmade, mirror image lakes, North and South Lake, known locally as key lakes because they resemble the shape of a house key from above. I drove under the causeway that goes to Ocean Boulevard, the beach and waterfront promenade, and walked along Seventh Avenue, which follows the Stranahan River and leads to the Hollywood City Marina. The wall looked a little run-down, and there was evidence of some puddles in the street. A few coconuts had washed into the nearby rock jetty, and other trash filled the cracks. At the marina, some serious-looking police were gassing up patrol boats. Instead of talking to them, I asked a person walking two small dogs if he walks there every day. Nicholas Tranchina said he did.

"Do you ever see the water come in over the wall?" I asked.

"Seems to come right through over there," he said, pointing to a wall under the bridge and by the playground. He took a seat on a bench. "Why?"

"I'm working on a project about climate change and sea level rise."

I never know what to expect at this point in the conversation. Am I friend or foe? He looked over at the officers working on boats then shielded his eyes from the morning sun and looked back at me.

"I used to fly Air Force Two for Al Gore."

I patted his dogs, urging him to continue. "I was in Nashville election night, celebrating. We thought he had won. Then someone came in and said put away the champagne."

Broward County was one of the places election officials were reviewing dimpled ballots, divining voter intent, before the recount was stopped. In the certified result, Bush won by 537 votes. According to the Florida Ballot Project of the National Opinion Research Center at the University of Chicago had the recount continued by most prevailing standards, including marks on punch cards, Gore would have won.

I drove away thinking of what might have been. And what might still be. "I flew on Air Force Two for eight years," Gore liked to say in talks he gave following the election, "and now I have to take off my shoes to get on an airplane." He leavened his message of climate change doom with humor, referring to himself as a "recovering politician." He also warned about a rise in temperature that was not linear, that it could take big jumps or accelerate.

The reason I was there was due to talking with Jennifer Jurado, the chief resilience officer for Broward County. Jurado told me about the lakes and seawall and that there was "very little capacity left in that system." She and the county she represents are part of the Southeast Florida Regional Climate Change Compact, which sets both the policy and tone for much of the adaptation work for southeastern Florida. Broward, Miami-Dade, Monroe, and Palm Beach Counties executed the compact to respond to the impacts of climate change and coordinate strategies across county lines. For example, they came together to use a uniform sea level projection as they were all using different maps.

A sea level rise work group of the compact recommends using three projection curves, the NOAA High Curve, the USACE High Curve, and a curve corresponding to the median of the IPCC Fifth Assessment scenario. Using these tools, the compact projects in the short term six to ten inches of sea level rise by 2030 and fourteen to twenty-six inches by 2060 (above the 1992 mean sea level). In the long term, sea level rise is projected to be between thirty-one to sixty-one inches by 2100 (from about three to five feet). For critical infrastructure projects, they recommend the NOAA High Curve with planning values of thirty-four inches in 2060 and eighty-one inches in 2100 (almost seven feet).

With a shared problem, they share resources and work products. "We're more powerful if talking with one voice." They helped defeat the 2015 solar amendment disguised as "solar choice," a fake referendum that would in fact place barriers to solar growth. Since their formation, other regional compacts have formed. In terms of population, the four-county compact is greater than that of thirty states. "This issue is larger than any of us individually," Jurado told me.

Two years after its founding, the compact released an action plan containing 110 items, from deploying clean energy to identifying vulnerabilities. One of them relates to the Hollywood lakes and installing flap gates or one-way valves. At high tide, water would "backflow" from the lakes up through drains and into neighborhoods. With the backflow device, rainwater flows down but tidal surge is prevented from coming back in. The city will install a dozen such devices in both North and South Lake neighborhoods. They have made many other adaptation measures in the plan, from planting trees to advancing renewables.

Patty Asseff is a realtor in the area who served as city commissioner and ran for mayor as a Republican. She is also the chair of Southeastern Clean Cities, drives a Tesla, and attended with Jurado the 2015 Rising Tides Summit, a bipartisan gathering of nearly forty U.S. mayors and local elected officials from eighteen of the twenty-three coastal U.S. states, learning a great deal about strides cities can make to become more resilient. Though she brought these issues to the fore during her candidacy, she lost the race. "No one seemed to care." She has seen flooding near her South Lake home, which is pretty much at sea level. And she worried about the effect much of the bad news about climate change had on her ability to sell real estate in South Florida. Since neither she nor many longtime residents want to leave, she supported practical measures that help her city prepare in a nonpartisan, just get-it-done matter. Forget debating the science, do something about what she and others can see. The problem is, implementing much of what they want to do, including catch basins, costs money, and "we don't have smack." But driving around the palm-lined streets, it looked like there was at least a little smack.

There appeared to be even more money as I drove into Coral Gables, where I was to meet with Mayor Jim Cason, who also attended the Rising Tides Summit. Billboards featured love handles disappearing, breasts enlarging, wrinkles smoothed over. "Look your best at the beach." We spied no fewer than five places for the surgically enhanced.

Cason is working to protect the city from some of the flooding it already experiences and to prepare it for more flooding that will most likely accompany rising sea levels. Though Governor Scott questions the cause and science, Cason and other Republican mayors in South Florida are making pragmatic decisions about the issue.

Coral Gables's city hall is made of native oolitic limestone. Climb the steps and walk over an ancient sea bed. There are fossils in the rock, mollusks and coral, and evidence of the porousness it is known for. The building is on the National Register of Historic Places for something known as Mediterranean revival. Denman Fink, once the art director of Coral Gables, had a hand in designing the building. A mural of his, *Marine Painting*, hangs in the stairway modeled for one in Cordova, Spain. In the bell tower high above, another of his depicts the four seasons.

While waiting for a meeting, I signed the guestbook. Ahead of me were consulate generals from France, Brazil, and Colombia. I felt in pretty good company and would soon learn the reason. Cason had been an ambassador to Paraguay and worked as a principal officer with Cuba.

The city of Coral Gables is one of the original planned cities in America, part of the City Beautiful movement that was inspired by Frederick Law Olmstead, who designed New York's Central Park. Wide tree-lined avenues, green space, ornate plazas, and public fountains were all visible. But to make the paradise among citrus groves, rivers were dredged, the fill used to make high ground. Now, as sea levels rise, all may not be well in the iconic community.

The first thing Cason said to me is that the city faces an existential threat. And despite Secretary of State Rex Tillerson saying climate change could be an engineering problem, Cason saw no obvious one for his low-lying city. At the Rising Tides Summit, "all had a solution except for us." Seawater incursions were beginning to affect water supply. He pointed at a new Lidar map of the city's elevations, much of it at about three feet above sea level. "All that brown goes."

"This is the dilemma of a mayor today," he said, showing me a cartoon of people waist deep in water complaining about traffic circles and trash. Cason said he had never received an email on sea level rise but he had on trash. He feared people were concerned about the daily problems but did not see the big picture. "This is not a partisan issue," he told me. Cason asked big questions as photos of his family and grandchildren flashed on a screen behind him. "How will history look back at you? Did you fail as a leader to discuss unpleasant things so people could make a proper determination of their risk?" If people were in danger of losing homes or property, he thought they should know. "You have to tell people so they can plan."

Cason said he was working to prepare for the day for when the reality comes. He talked about $3.5 billion in property values, but we "don't need more tax cuts. We need twenty million to build seawalls." Coral Gables has some forty-seven miles of coastline and much of the tax base is near the water. He worried that if some of those homes were affected—and they can go for $20 million—"our ability to maintain services and quality of life plummets. You start spiraling down from there."

But it's not an issue talked about in the exclusive, gated communities where $50,000 is needed just to apply to live there. Those people pay cash for homes, but he could see a future when those without adequate cash would have trouble obtaining a thirty-year mortgage or flood insurance. He mentioned that the city has some ten thousand septic tanks. If the seawater floods those, "the you know what is gonna hit the fan—and bubble into the streets."

The psychologist Abraham Maslow proposed the theory on the hierarchy of needs. The most fundamental needs are taken care of first, such as those necessary for survival. Wealthier communities seem to take the lead on much of climate change adaptation planning. Poorer ones in South Florida—Hialeah, Homestead, North Miami, Florida City—have other issues to attend to, such as crime or homelessness. But Cason could see a time when doing something about sea level rise will affect those other needs. "We are nice to each other now but maybe not when things go bad."

Mayor Cason showed me a picture of early Coral Gables. The Biltmore stands proudly in the background, circa late 1920s. A gondolier guides a boat under a bridge on a canal that used to go all the way to Tahiti Beach on the coast. Cason calls it the "before and after."

Before I left, he showed me a poem he had altered, Robert Frost's "Fire and Ice." "Some say the world will end in fire," the poem begins, but "some say in ice." The poem brings us into the middle of an argument about whether the world will burn or freeze. But it carries another, metaphorical meaning. If fire is desire and ice hate, those could also do the job, especially ice/hate as a kind of silent killer, which could come at that time Cason envisions when we are not so "nice to each other," his downward spiral. Cason edited the last few lines to insert how "lack of" ice could cause destruction because it floods his city. Driving around South Florida, I thought of a third possibility for how the world will end—in traffic.

The melting of glaciers is an area of expertise of Hal Wanless, chair of the University of Miami's Geological Sciences Division. Wanless, who is in his seventies, has spent half a century studying geologic change in South Florida and elsewhere, including the poles. Said Cason, "I never heard of sea level rise until I talked to Hal."

I joined him at Burger Bob's, classic diner fare and the "tenth hole" of the nine-hole city-owned Granada Golf Course. When he looks at the data, Wanless foresees the high end of sea level rise projections. Before we even started talking about numbers, he mentioned relocation, how it would be more economical to move people than to rebuild infrastructure. He talked of cleaning the land, the industrial areas that will be inundated. His mind seemed to work like the Sea Level Rise Viewer map from NOAA: slide the scale and see what floods. But that map only goes to six feet. Wanless's would go higher.

He worried about the heat built up in the ocean. Over half of the excess global warming accumulated in the oceans has occurred since 1997, so "you can't just blame the old fogies," he said with a wink, turning toward Sam. And that heat will start melting ice. Most of the models project a gradual acceleration. But when he looks at how the ice melted in the past, there were "rapid pulses of rise . . . fast enough to leave drowned reefs, sandy barrier islands, tidal inlet deltas, and other coastal deposits abandoned across the continental shelf," he wrote in a policy paper. When the sea was some 420 feet lower, these reefs and mud flats topped, saw no slow accretion. They were drowned rapidly as an ice sheet disintegrated.

When ice melts, there are accelerated feedbacks. Water on the surface of the ice absorbs more heat. This heat melts more of the surface. The melt then percolates down through the ice, "lubricating" the base, permitting faster motion, fracturing. Wanless had the chance to fly over the Greenland ice sheet and witnessed a deep channel some five hundred feet below the surface. One of the things Wanless took away from climatologist James Hansen, formerly of NASA, was that ice melt is nonlinear.

The problem with the existing projections? Wanless said they were premature not wrong. "Most of the IPCC stuff is 10 years old," reflecting the time to gather the data and the time to publish it. He thought that with the evidence we have now, projections should be around ten to thirty feet of rise by the end of the century. Less than 10 percent of land in South Florida is above ten feet. Margaret Davidson, NOAA's chief resilience officer, told a 2016 meeting of insurance and risk management experts in California that new information indicates nine feet by midcentury. "The latest field data out of West Antarctic is kind of an OMG thing," she told the crowd, marking the first time in history a senior scientist has used the abbreviation—almost exclusively uttered by texting teenagers to show surprise—in the presence of insurance executives.

During our time in South Florida, there were unusually warm temperatures around the North Pole, 2.5°C above normal.

When our conversation turned to politics, Wanless expressed the view that "I think Trump understands. He played the game to get elected." The president's cabinet picks, however, say otherwise. "His daughter will lead him into it." Early that morning, Wanless looked up the elevation of Mar-a-Lago. The "winter White House" will have water at the door at six feet. Three feet would turn Trump Hollywood, one of his South Florida properties, into an island, swamping most of the neighborhood. At six feet, a boat would be useful to get to the door. At that level Doral, his golf course ten miles inland, starts becoming amphibious, crocodiles teed off.

Wanless and nine other Florida scientists sent a letter in December to Mar-a-Lago requesting a meeting with Trump on climate change. "Many of Florida's waterfront properties (including yours) are vulnerable to even minor increases in sea level because of erosion and storm surge," they wrote. "This is not a distant threat. Climate change is making an impact today."

What Wanless described was a little scary to contemplate. "You're going to see interesting changes in your life" he said to my son as we finished our lunch. Talking with Wanless, it became clearer why Mayor Cason was such a convert to climate change, planning for his city's future. "Young people should realize the opportunities." That involves investing in renewables and setting aside natural resources, groundwater. "If you get this, and plan for it, things will be alright." If not? "It's gonna be chaos."

Phil Stoddard, the mayor of South Miami, says "eventually we are going to be leaving." He envisioned a schedule to take land lying at two feet, then three, and so on, turning them into the protective wetlands of some kind. Stoddard told writers Stan and Paul Cox that since Americans do not like "retreat," he preferred "rolling easements."

For decades, Wanless and others scientists were met with skepticism when they discussed sea level problems, modern-day Cassandras. Recently, however, as more and more residents have found themselves wading through water, people have been listening. A headline waving "Goodbye, Miami" in a 2013 *Rolling Stone* article that featured Wanless captured the attention of both the real estate investors and the tourism industry. About that article

Jennifer Jurado told me that at the time, "there was a much different conversation on the ground."

As we left, we had a decision to make. Should we visit Miami Beach, a low barrier island with world-class beaches and nightlife, the Art Deco playground? On the radio, we heard that the city was contemplating moving last call three hours back to 5 a.m. The area floods regularly so it was worth checking out for my purposes. Miami Beach is investing $400–500 million to elevate roads and install about sixty pumps that could capture water and push it into Biscayne Bay. But if you raise the road, the water has to go somewhere, and it could end up in the homes or basements yet to be raised. A month ago, during a king tide, an octopus was washed into a Miami Beach parking garage. Pictures of it splayed on the concrete floor led to claims it was bogus, brought in from the market, but biologists confirmed that it was a local species all too real, even if the circumstances of an octopus occupying someone's parking space were beyond strange.

In Florida, it may not be the elephant in the room we need to discuss but the octopus in the basement.

Wanless told me that at one point the city had pumps that could only deal with fresh water or that were not high enough to handle being flooded. Those have been upgraded and some of the pumps are mobile, tanker trucks that can pull up to a flood and suck up the water. Chris Bergh, an adaptation specialist with the Florida Keys Nature Conservancy, told me that pumps are "only gonna get you so far before you're pumping sea back into the sea."

Instead of visiting South Beach we opted to take a drive out to the Everglades. On the way out, we stopped at a spot behind a rental car place near the airport that Wanless had told us about. Where highway ramps converged and planes flew overhead, just by the club Pink Pussycat, was one of Florida's many water control structures. A boom held back trash and a gaggle of white ibises picked through the grass on the far side of the canal.

This barrier, several miles inland, is the boundary between the fresh water coming in from the Everglades and the salty water intruding from the sea. When they drained the Everglades, water levels dropped by some seven feet. At gates like this, fresh water can be let out during a flood and the seawater held back. This freshwater "head" had a foot and a half when built in the 1950s and helped keep pressure in the aquifer and kept the salt water at bay. But increasingly, the seawater is encroaching on the height of the gate. Expensive pumps, at $70 million a pop, have to be installed because gravity no longer works.

Jayantha "Obey" Obeysekera, a hydrologist with the South Florida Management District, told me more about the water control structures and their function as water supply. Obey talked about the porous, Swiss cheese limestone underground. The sunshine state averages sixty inches of rain per year so you would not think they would have a problem with water. But seawater threatens it because of that limestone. "We get water from four different sides," he told me: from both sides of the peninsula, from above and from below. "We can't build Dutch-style levees here or surround ourselves like New Orleans. The water comes up from below." Bruce Mowry, Miami Beach's city engineer, told writer Elizabeth Kolbert that he would like to lift the city up and put an impermeable membrane underneath, like a tarp under a tent. He was also intrigued by injecting resin or clay into the limestone, plugging the chinks in the rock.

Seawater pushes the fresh water to the top of the aquifer, but if it pushes too much, it goes beyond the depth where local wells draw from (fresh water has a lower density than salt water). That has already happened in parts of Broward County, where the city has had to drill for water further inland. But cities in South Florida cannot go too far west or they bump into the Everglades. And if the Everglades dries up more than it already has, peat soil that provides the filtration and structure for an entire freshwater ecosystem—and much of the region's fresh water—could collapse. To handle the drinking water problem, South Florida is relying more and more on desalination, a complex, expensive, and energy-intensive process that removes salt from the water.

On the way west out of town, we drove through the town of Sweetwater, between four and five feet of elevation. Under the city's streets and lawns runs the eastward flow of the fresh water from the Everglades and the town floods with every rain. Many have argued that the Corps of Engineers should raise the water level of the Everglades, increasing the eastward flow pushing back against the seawater. But that spells trouble for Sweetwater, which is 96 percent Latino with a median household income of $34,000. Towns like it, with little financial or political capital, could be among the first to be surrendered. How compensation will work, and where residents will go, is yet unknown. The affluent will absorb the uninsured losses, especially of second homes, and move or bunker up. The poor and middle class, with life savings tied up in homes, could take the biggest hit, facing the reality of climate change directly.

We drove out the Tamiami Trail through the Everglades, which writer Marjory Stoneman Douglas called a "river of grass." The section we traveled through had some of that grass but also shrubs near the road, clumps of sawgrass, and an occasional hammock of trees. The Miccosukee tribe operates many of the airboats and a casino. Before Europeans arrived, they lived in Georgia and Tennessee but were chased further and further south until they ended up on a few hundred wet square miles between Miami and Naples.

We had been looking for a bar where we could watch an English Premier League Soccer game Sam wanted to see. We figured that with a large international population, we would have no trouble in Miami. But we did not find one. Instead, he streamed the game through his phone and the car speaker.

I turned onto Loop Road, which leaves the Everglades and enters Big Cypress National Preserve. We stopped before we left the pavement because we were traveling out of cell range. I parked the car near a small bridge and got out to look at the water beneath. On one side, a young heron speared minnows in the shallows, not affected by my presence. On the other, I heard squawks, flaps, and a low moan, mingled with the faint sound of an English broadcaster and the cheers of the crowd, as if they were rooting for the fowl. I saw an alligator too, then another, but neither seemed to be moving at the pace of the game.

During this trip, we had learned on the radio that the word of the year was *surreal*. Not mentioned by Merriam-Webster lexicographer Peter Sokolowski in relation to surreal were the Cubs' World Series win, David Bowie's and Prince's unexpected deaths, and the massacre at Orlando's Pulse nightclub, all of which fit the definition: "Marked by the intense irrational reality of a dream." It described the scene before me.

After the game, we kept driving through dusk. At each small bridge we passed, we scared up a flock of birds. Some flew down the road in front of us, a game we were used to playing in a canoe. They fly downstream until we reappear, then take off and fly further down again. At several spots, we stopped just to take it in.

Cypress trees are barrel shaped at the bottom, where they send their water-loving roots through the water's shallow expanse. These roots can break through the limestone bedrock and take hold to grow. In the fading light and whispered dark of the swamp, I had never seen a wonder quite like it. The place thrummed with wildlife. It awed, soothed, and humbled—a mythic vastness, primordial ebb and flow, just an hour west of a major city.

We kept driving as night fell, searching the sides of the road for the eyeshine of what we hoped would be a panther. We did see a bird of some kind, a small hawk we thought, that kept diving at us like it would attack the windshield. We figured it was catching insects that were attracted to the headlights.

As it grew later and later, we finally saw a snake crossing the road, one of the reasons my son came on this journey. I stopped the car and he inspected. It was a cottonmouth. Sam watched it for a while—me too from the safe vantage point of the car. Water moccasins will come out at night and use a kind of thermal imaging to detect warm-blooded prey. Sam tried to shoo it off the road with a stick. It did not respond to the first prod but then sped off, undulating across the sandy road as if swimming. I have seen friends I hike and fish with do a kind of vertical leap when they see a snake, as if they are trying to go up because they cannot move away. Sam also jumped when the snake took off, but it was a joy to be in the presence of such a complex, strangely beautiful creature. Better than a Liverpool goal.

He wanted to do the loop all over again. Since we did not know when we would be back, it made sense. Also, what would the state of the Everglades be in the future? One of the most remarkable ecosystems on the world is threatened by the same features that threaten the rest of South Florida—it is flat and low-lying. Much of the water that feeds the park is diverted and what flows in is polluted with nutrients from fertilizers. At the southern tip, the water is becoming salty, mangroves are moving in. The Comprehensive Everglades Restoration Plan (CERP) would "restore" the Everglades, but it may just halt the damage for a while. There may be no turning around. Eventually, in a race of an ocean tide pushing in and freshwater river flowing out, the processes set in place by global climate change will put much of the Everglades, and South Florida, underwater.

To see what those changes were beginning to look like, I met Chris Bergh of the Florida Keys Nature Conservancy. Chris said the keys have seen "dramatic changes" driven by the nine inches of sea level rise in the last century. For him, things began to crystalize around ten years ago. There was Hurricane Wilma, Al Gore's Inconvenient Truth, and the IPCC's Fourth Assessment. He produced a report, "Sea Level Rise on the Florida Keys," which looked at different scenarios, at where new shoreline would exist and what would property values be. At the high end, four and a half feet, some 95 percent of the island would be affected at a loss of $1.6 billion in property.

When the report came out in 2009, he was afraid to put his name on it. He was concerned people he grew up with, those in the construction trades and in real estate, "would hate my guts." But as he met with people and spoke at county commission meetings, people who were paying attention saw what was going on. On the keys, "people are tied into the tides."

Bergh, who lives on Big Pine Key with his family, is at work on another adaption plan for endangered species. One of the things he looks at is what to do about Key deer. These deer are a subspecies of white-tailed deer, only much smaller. They migrated across a land bridge during the last glaciation, but then the water rose, and they adapted in place. There were as few as fifty deer left in the 1950s when they were protected by law. The population increased to about one thousand in recent years and is now about eight hundred in the wake of a screwworm outbreak, an aggressive pest.

One of the first things he said to me was that they may need to practice *ex situ* conservation, which is the practice of preserving a species outside their habitat, but if you move them where there are other deer, "their genetic distinction would be swamped."

The deer that live there and the pines that give the island its name depend on a freshwater lens in the rock matrix, surface fresh water, which makes the lower keys biologically unique. As Chris explained it, the upper keys are made of Key Largo limestone, which are fossilized coral reefs. This is very porous, a sieve. The lower keys consist of a layer of oolitic limestone on top of the other. It's not as porous and holds water like a sponge, which can be bad when salt water comes in on top. In Big Pine Key, there is no viable way to prevent seawater from seeping through the porous geology and contaminating the underground freshwater lens that feeds the pine forests and deer. The pine and deer are tapping into the lens, but as it shrinks, and as salt water pushes against it, they are dying out. The pines and the deer are somewhat codependent. Pine indicates the presence of the lens and their flammable leaves are a big contributor to the fire that preserves the dominance of the grasses, forbs and shrubs deer like to eat, and keeps them from growing out of their reach. Controlled burns may be one of the adaptation strategies.

Floods, fires, and pestilence—it was beginning to seem biblical.

We ate lunch at the No Name Saloon, dollars tacked to the walls, a former brothel. Over a Cuban sandwich, he listed off some of the other species threatened by sea level rise: "marsh rabbit, Key Largo woodrat, Key Largo cottonmouth—how much ink do you have?"

Chris was also worried about the coral reefs, another economic driver. Less than 10 percent of the reef system is now covered with living coral. The reefs experienced back-to-back major bleaching events in 2014 and 2015, mainly due to heat stress. Scientists anticipate that as early as 2020, the reef could be in line for almost yearly bleaching events, in some cases killing them.

After lunch, Chris took us to a mangrove marsh, one that is super salt tolerant near the western shore of the island. Off in the distance we could see pines but some were dying. We were looking at the ascending tree line of salt-tolerant mangroves and then a little up, the buttonwood, then some scrubby woods and finally the tall pines, though a few looked dead. There, the transition from fresh water to salt was happening.

Chris wanted to show us that change up close. As we walked a trail, he pointed out poisonwood, related to poison ivy and important wildlife food with super fatty berries birds love. It should burn periodically; otherwise other species take over. He showed us where wildlife could access fresh water, down in a hole.

Finally, we entered a kind of no-man's-land. There was a utility pole because at one time the area was platted to be subdivided before being deemed too ecologically sensitive. Chris pointed to a ditch dug for mosquito control that accelerated the transition to marsh. Freshwater could exit faster but salt water could enter. It was a moonscape, a rock barren, and he said there was a lot of it in the keys. The red mangrove was coming in and thriving and the only indicator that this was once a forest—a pine stump. Spring or king tides bring salt water in and trap it in the rock, creating a hypersaline environment. The "king tide bites a little deeper each time. Aerial photos over time show it. Now you see the vegetation. Now you don't."

Chris's house was a few streets over. When I asked him if he thought about moving, he said he is building an addition. However, he assumed that the value of his property will go down. He wanted to keep living here, as much of the community does, prospering economically. He thought adaptation would be the key. Some people already live on boats. "Before it's time to split, there will be a lot trial and error. Stainless steel cars, junker cars, who knows, little things. People will find a way."

In Key West, they are raising the building height cap. They made the ordinance because they did not want Miami Beach high-rises and because of big winds they can have during storms. But homes that are built say four feet above ground go up against the thirty-five-foot cap. In the absence of national policy, much of the need to adapt to sea level rise will fall to the localities. In the coming years, Chris anticipated pilot projects and even private individuals leading the way. However, he tempered his optimism: "We could stop burning fossil fuels, I'm still screwed. So are my nature preserves. So we have to adapt." The point for him is to plan for the change ahead and adapt while there is still time.

On the way out, as if on cue, we saw a Key deer nibbling at a neighbor's vegetation. In the industrial era, canaries warned miners about dangerous air in the mine. Today, small deer—fewer than eight hundred of them—roam Big Pine Key, searching for fresh water. The eventual *ex situ* adaptation plan for Key deer (*Odocoileus virginianus clavium*) presages the relocation of Key people (*Homo sapiens floridianus).*

Since we had come this far, like many travelers, we had no choice but to head to the bottom of it all, Key West and its cultural, artsy, free-wheeling scene. We nabbed a parking place next to the "rolling reefer," a truck decorated with shells, seaweed, some rubber sharks and a ray, a few skulls: the words "some art rocks—some rolls" were painted on the side. We walked around to Ernest Hemingway's house. The wealthy Uncle Gus of his second wife, Pauline, helped Hemingway buy the home as a wedding present in 1931. He purchased it for $8,000 in back taxes. At sixteen feet above sea level, it is the second-highest place in Key West (Solares Hill is the highest at eighteen, unless you count Stock Island Landfill at sixty-six feet).

I looked for people to talk to about their thoughts on climate change, but many I overheard were not speaking English. It seems many international visitors choose Key West as a destination, though some were on a work visa. Others had young children, which seemed odd for a party town, and I did not want to disturb their souvenir hunting.

Instead of touring Hemingway's house, which had a long line of tourists, we ended up in his former haunt, the legendary Sloppy Joe's Bar. Hemingway himself helped come up with the name. He knew of a Havana club that sold liquor and iced seafood. Because the floor was always wet from melting ice, patrons taunted owner Joe with running a sloppy place. We ordered mojitos and a slice of Key lime pie.

Hemingway wrote at a time that predates widespread concern over human-made climate change. However, his description of the glacier in "The Snows of Kilimanjaro," a story he wrote in Key West, has been invoked to draw attention to the problem. Hemingway wrote about the "unbelievably white" ice fields on the "square top" of Kilimanjaro, from Swahili words for "shining mountain." In 2003, a group of scientists published a study based on their research of drilling ice cores and comparing aerial photographs since Hemingway's time and before to show that the ice field had decreased by 80 percent. By 2020, they predicted, no more shining on the mountain.

Al Gore ran with the information in *An Inconvenient Truth* (2006), but other research has shown that the ice is disappearing more because of a long-term change in the region's hydrology and solar radiation than warming. In fact, temperatures on top of the mountain have stayed fairly constant. Patrick Michaels, a prominent climate change contrarian, took up this later research in his newsletter for the Greening Earth Society, now defunct but with links to the fossil fuel industry. Though melting may not be the cause of the disappearing ice, climate change may still play a role as the warming of the Indian Ocean may affect changes in the east African climate. A shortage of precipitation that is a result of climate change could be causing increased rates of sublimation, or the conversion of ice into water vapor. As Andrew Revkin, formerly of the *New York Times* Dot Earth blog asserted, Kilimanjaro has become an icon of climate change debate rather than climate change itself. Still, the glaciers are melting the world over because of warming, and that ice has to go somewhere. It may eventually wind up in the floor of a bar.

I started asking people inside what they thought. One stocky gentlemen at the bar told me no way the planet is warming. "It was too cold last winter. That's why I'm in Florida." I asked our server, Molly, who moved to Key West from central New York two years ago what she thought about climate change on the keys. "People just accept storms and floods as part of life," she told me. "They just board their houses up and ride it out. They're a lot less fussy than they are in Miami. If the flood comes, they'll throw a party."

Alex, a longtime local, told me that at some point his house will be impossible to sell. "I could get a chunk of change for it right now. Maybe I should sell it and move to the hills . . . to Virginia, where you live. But it's just so beautiful here." "This is just a wonderful place to live," he said, but he slurred these words some. "We're all going to enjoy it as long as we can." We clinked glasses to that. I feared that Alex was luring me in to buy him a drink. The formula: find out where the tourist is from, say something nice about that place, make him feel like they belong in Key West. Let's drink to that.

What Alex said did not seem far off to what many told me. They knew a time was coming, but they did not want to leave. When I asked Hal Wanless why he thought his message of catastrophic sea level rise had a hard time getting through, the geologist grew philosophical. "There's a hunger for a permanence of place deep in our being." At the end of *Oh Florida*, a book on how Florida's "weirdness" influences the rest of the country, journalist Craig Pittman writes that "Florida will be a personal paradise, yours to own as soon as we fill in this hellish swamp. Florida will be growing in perpetuity, so long as we keep persuading suckers to move here. Florida will be underwater soon—let's just hope it's not as soon as experts predict."

If Alex was trying to ply me for a free drink, it worked. But halfway through that one, he grew a little surly, resigned. "What can we do anyway? What one thing could we do realistically right now that would do anything to reverse what's going to happen?"

We left Alex in the bar to go set up camp at Bahia Honda State Park. We had time to swim just as the sun set, slicing through a gap in the old railroad bridge, replaced by Seven Mile Bridge. It felt great to have water on the skin, the color of it calming, relaxing the mind, but the sun fading over rusted, decaying infrastructure felt like a glimpse into some time capsule, the before and after Jim Cason had talked about.

We made one more stop on the way out of Florida. The next morning, we woke early and drove to Vero Beach to see Steve Traxler, a senior biologist with the U.S. Fish and Wildlife Service. Steve is working on the adaption plan with Chris Bergh. He took us to Pelican Island National Wildlife Refuge, the first ever, created in 1903. When Theodore Roosevelt visited then, there were some five thousand brown pelican nests on the tiny island we looked out on. It did not seem possible, but it struck something in him.

At the turn of the century, the Audubon Society had formed and many local chapters, headed by women such as Harriet Hemenway of Boston, were outraged by the slaughter of birds for their plumage. Florida species were especially valued for their use as decorative plumes in millinery. By 1900, some five million birds were being killed each year to make hats, decimating shore bird populations.

When it was discovered that the island was the last breeding ground for brown pelicans on the East Coast of Florida, the Florida Audubon Society hired a local homesteader and German immigrant, Paul Kroegel, to do what he was already doing—protecting the birds from hunters. Kroegel became the first wildlife refuge manager, hired for a dollar a month.

The boards of Centennial Trail list every National Wildlife Refuge created in reverse chronological order and lead to an observation deck. From there, we spotted a few pelicans and Steve told us about the Indian River Lagoon we looked out on, home not only to birds but some eight hundred types of fish, where temperate and tropical zones overlap. As temperature changes, what will happen?

We left Steve thinking about that farsighted investment over a hundred years ago to set aside some wild places both for species and because of our deep-seated need for them. At the turn of the last century, when Americans had still not finished taming every bit of land they could, some set aside land to be left alone, refuges. They were an investment of sorts, an asset entrusted to the care of later generations. And we thought about how important were such places of refuge, inviolate, separate from the noise and hum. Marjorie Stoneham Douglas used to say she fought for the Everglades because it was a test. "If we pass, we may get to keep the planet." Some places, like Florida, or Alaska on the far diagonal shore, can seem too large for damage, but it is there. It is most definitely there.

Needing to head for home, we drove on. We stopped at dusk to grab some dinner. I was tired and ready to find a place to sleep but Sam said he would take the wheel. I snoozed in the passenger seat. Before I knew it, he was waking me up as we passed downtown Charlotte, North Carolina. Lawmakers had failed to repeal the "bathroom bill" that required people to use the restroom that matched the sex on their birth certificate regardless of gender identity. Downtown buildings were lit up in rainbow.

It was after midnight, my birthday, and I had passed the midcentury mark. Two hours later, we would pull into our driveway. I slept again only to wake on the windy road that leads to our house as Sam hit the brakes. A white-tailed deer, twice the size of those Key deer, leapt the road in front of us, the fence next to it, seeming to bound the creek beside it, disappearing when out of our lights like some mythic, winged horse. Normally, after a long trip, I am glad to be home but part of me was still in the Everglades, back in the keys. Just the day before, I was in a bar in Key West and I was half there again, wanting to finish my drink with Alex. He was right. It was a beautiful place. But I wanted to tell him there was something we could do, little things each and every day, big ones too, if we plan for what is ahead. I shook my head as if to get awake. Sitting there in the dark, the engine ticking as it cooled, I half wondered if I was here or there. Had any of it really happened? I was glad a young person was at the wheel, while us old folks were lost in a surreal dream. When we woke up, we barely recognized the place before us. It had all changed. And we might have done something about it.

7

Springing Back

Resiliency on the Gulf Coast, Louisiana and Texas

I went to New Orleans because I wanted to see how the city was faring more than a decade after Katrina devastated the city, a cultural treasure home to some four hundred thousand people. I wanted to learn more about how they rebuilt the system of levees that failed and if they were stronger now and ready for the next event. Flood risk due to sea level rise will grow more severe for New Orleans, already below sea level and sinking or subsiding. Was the city ready for new storms, as temperatures rise and as the conditions that created Katrina get worse, making hurricanes more intense?

But it was spring break for me and three fellow travelers, all of them in their first year of college. I wanted to survey some of the past (and potential) destruction. They wanted to stroll Bourbon Street and then canoe and camp on one of the nearby bayous, have a good time. It's a paradox embodied in the region's music. Even in the face of oppression, violence, heartache, there's a song to sing.

With me was my son Sam and his friends Dean and Cameron. They have known each other since elementary school. I used to coach them in youth soccer, bring them on hikes. Now, they were glad for a ride to the Crescent City.

But they ditched me as soon as we got to town.

When I set out from my hotel, I hoped to hear some music but the streets seemed empty. It was Sunday night and a week after Mardi Gras, and I thought perhaps I had entered a pause in the fabled mirth and merry making. On my walk I passed one of the "evacuspots," a tall metal stick-figure sculpture waving as if to flag down a taxi. They are official meeting places for anyone needing a ride out of town during an evacuation and a reminder of the floodwaters that trapped residents, many of them in "shelters of last resort" such as the Superdome or Convention Center. The one I passed stands in front of the historic Congo Square, where slaves could gather on Sundays to set up markets and play music.

As I rounded a corner in the French Quarter, the streets came alive. After oysters and a beer at Felix's, I walked some more down the cobbled streets, which had an odor of revelry, alcohol and seafood but something else too, maybe days-old trash or upchuck. Sure enough, a street performer was playing "Let the Good Times Roll," written by Sam Thread and his wife, Fleecie Moore, but recorded by Louis Jordan in 1946, Ray Charles in 1959, and B. B King in 1999, among others. Let's have some fun because "you only live but once, and when you're dead you're done." The subtext of that "good time" is the ever-present threat of destruction lurking below the surface. During Mardi Gras, the expression is "laissez les bons temps rouler."

Perhaps Katrina exposed the fallacy of the city's "good time" ethos, as some of the neighborhoods where the festival traditions are rooted were largely destroyed. Yet that tradition is precisely where the city seeks its inspiration for renewal. At a "brass band funeral," only dirges are permitted before the body is "cut loose" and the jazz begins. In *Why New Orleans Matters* (2008), writer and musician Tom Piazza sees the "jazz funeral" as symbolic of the resilience of New Orleans culture and its people. The jazz funeral reveals that catharsis through music is a kind of coping mechanism, a function of life in the city's poorest neighborhoods. "And an attitude towards life that includes a spiritual resilience which has spoken to people around the world for a couple of hundred years."

After the winds and waters of Katrina struck in 2005, 80 percent of the city flooded. The storm killed more than 1,800 people, did $75 billion worth of damage to the city, and left more than 100,000 homeless. Some describe those who left and never came back as among the first climate change refugees. While Katrina was more than ten years ago, it still serves as the model for what could happen as storm intensity increases and sea level rises. It is also an event that city officials, policy experts, emergency responders, and insurance companies study to determine what went wrong and how to prepare for the next disaster.

Resilience is the word they embrace in New Orleans: the capacity to recover, bounce back, be even stronger than ever. The word lends itself to the psychological, and New Orleanians experienced their share of trauma from Katrina. I asked one resident of Lakeview at a lunch counter serving red beans and rice in Bucktown if she still felt vulnerable years later. "Katrina was so traumatic it's hard not to feel vulnerable." Resilience, according to *Psychology Today*, is that "ineffable quality that allows some people to be knocked down by life and come back stronger than ever."

I was there in part because there was a conference in town devoted to resilience, Res/Con. The conference debuted in 2012 as the International Disaster Conference, "the premier event for global disaster management professionals to discuss policy and best practices in preparing for and managing catastrophic events." It relaunched in 2016 under the current name with the banner "Adapt. Thrive. Sustain."

Since I had been focused on adaptation, I wanted to find out more about resilience and why it was the new term. Like one of the words in the conference motto, *sustainability*, I figured it was simply a way to cast a wider net. In universities and businesses, sustainability, which grew out of the environmental movement, morphed into a catchall for all things "socially good." Major retailers, like Coca-Cola and Pepsi, promote their commitment to "sustainability," and Res/Con was sponsored by a number of businesses but prominently by the Rockefeller Foundation, which sponsors the 100 Resilient Cities initiative, a global network of cities, including Norfolk and New Orleans, dedicated to urban resilience. Each city has a chief resilience officer who coordinates efforts. Many of these are directed at preparing for the effects of climate change, yet "resilience" seems to be another way to address them. It seems like basic urban planning and governance with the heightened awareness of the disruptions that can occur because of climate change.

Of course, John D. Rockefeller founded Standard Oil (later broken up into ExxonMobil among others). The burning of oil created the conditions that these cities now respond to. So to a skeptic, that the Rockefeller Foundation was a major player in climate change planning could look like they were covering tracks, atoning for the damage done.

Before the governor and the mayor spoke, I caught up with Barbara Morgan, a conference organizer (and former communications director to

Anthony Weiner). She compared resilience to education, how there is no one approach to student (or city) success. Indeed, it is a hot topic there too, as a quality to instill in children, though the recent import seems to be "grit." Mark Davis, the director of the Tulane Institute on Water Resources Law and Policy, told me resilience was a "little like heaven in that it can be anything you want it to be." But he said resilience tries to create some notion of what the outcome might be in terms of a better, stronger community.

Sam Carter, managing director at the Rockefeller Foundation, welcomed us and defined resilience: "The capacity of individuals, communities and systems to survive, adapt and grow in the face of shocks and stresses." Carter stressed *capacity* and said that it was not simply about flood protection but about the people in need and those doing the work to help them and others.

The word *resiliency* was barely in the New Orleans lexicon before Hurricane Katrina, according to Mayor Mitch Landrieu. But today, it is nearly inextricable from his city's initiatives. Landrieu was lieutenant governor when Hurricanes Katrina and Rita hit Louisiana in 2005. He recalled how his staff morphed from what was basically a culture of tourism promotion to one focused on emergency response and recovery. "How do you prepare for what you know is coming and for what you don't yet know about? How do you rebuild?" are questions Landrieu said government and business leaders should be asking. A resiliency strategy covers these scenarios, and the lack of one becomes apparent when risks emerge out of nowhere.

Landrieu recalled an incident that happened in 1972 when his father was mayor. Mark Essex, a Vietnam veteran, doused curtains of the Howard Jones with flammable materials and caught them on fire. Then he began to shoot firefighters. The mayor said New Orleans had experienced its share of disaster, including Katrina, but also Rita, Ike, Gustav, and the BP Oil spill. He said when Sandy hit, the Northeast woke up, but he thanked New York–based Rockefeller for being with them since Katrina. For him, resilience was basically the "theory of how to be stronger when something comes your way" but that no two attacks are alike. Still, people could be ready to respond, trained.

Gov. John Bel Edwards also spoke. He mentioned the historic flooding of the past summer and tornadoes that swept through eastern New Orleans a month before. With all that New Orleans had experienced, he wanted to highlight how Louisiana was leading in planning and response: "The knowledge and expertise developed here will be exportable." He also emphasized that there was "no time to waste," as floods and hurricanes will occur.

In his address, he touted the work of the Louisiana Coastal Protection and Restoration Authority (CPRA), a governmental authority created in the aftermath of Katrina and ordered by Congress. The 2017 Coastal Master Plan would soon be presented to them. He said it was a science-driven plan. It will spend $50 billion over fifty years.

Edwards also announced the launch of LA SAFE, Louisiana's Strategic Adaptations for Future Environments. LA SAFE will complement the state's Coastal Master Plan and provide planning for communities that will face resettlement due to land loss and sea level rise. He said worst case scenarios are becoming more and more likely, so there was a race against time in Louisiana, and "sense of urgency about coastal restoration."

HUD has provided a $92.6 million award, of which $40 million will be for LA SAFE. Nearly $50 million is reserved for the relocation of the residents of the Isle de Jean Charles and what is hoped to be "a national model for resettlement." A picture of the island flashed on the screen. He said the challenges there and in other parishes are among the greatest ever faced in the history of his state. The planning process, he assured us, would be inclusive, addressing literacy rates, language barriers, and the technical nature of information. Residents would "be involved in designing the methods by which they would like to adapt."

He reminded us that the need for resilience does not stop on the coast. During the floods of last March and August 2016, fifty-six of sixty-four parishes were declared disaster areas, some 85 percent of the state's population. Edwards reminded us of the importance of the role Louisiana plays in the national economy in terms of fisheries, navigation, and hydrocarbons. So it is important that we "get this right, the sea level isn't rising just here." And after a grim reminder, he told us that even though Mardi Gras was over, "to go out there and have a good time."

Before attending the conference, I met Mark Davis of Tulane's Water Institute. Davis told me that "teasing out what is possible from just boosterism is part of the adventure" in these discussions. In a sense, boosterism and denial are opposite sides of the same coin. Both rely on rather vague ideas and slogans.

But Davis said he can understand both. People are boosters because New Orleans is an important city. It is one of the country's top producers of seafood and energy, and its ports facilitate 20 percent of waterborne commerce. There is denial because there are no good options for some that may have to face relocation. But if current trends continue, and the same level of response, New Orleans may cease to be recognizable.

The master plan the governor alluded to begins with some sobering, honest news. Louisiana is losing land. Since 1932, the state has lost more than 1,800 square miles. The culprits include hurricanes, climate change, "disconnecting the Mississippi river from coastal marshes," subsidence, and "human impacts." The report spells out some of those, including how dredging canals for energy exploration took a toll on the landscape, altering wetland hydrology. Navigation canals allowed ships in but salt water too, degrading wetland soils.

After a flood in 1927, Congress authorized the Mississippi River and Tributaries project. The federal government assumed the entire cost of erecting levees, spillways, and other structures from Cairo, Illinois, south. After the completion, the Corps bragged, "We harnessed it, straightened it, regularized it, shackled it." But in preventing the flood, it also starved the barrier islands and delta.

Some three hundred square miles of marsh were lost between 2004 and 2008 due to Hurricanes Katrina and Rita. Increasingly, they are recognizing in New Orleans that they need a swamp between themselves and the sea. In the past, the concern was flooding from the river above so the levee systems were built. Now, it more likely comes from the gulf.

While resilience was the word for the conference, springing back, the soil in the area is doing the opposite, further sinking. The subsidence is due in part to extractive activity. When you remove fluids, you remove what props the land up. Some of it is natural, as delta soils compact. But the Mississippi has been relentlessly reworked with dams and levees, depriving coastal ecosystems of fresh water and replenishing alluvial sediment. Said Davis, referring to the city, "We sank it."

John Lopez, director of coastal sustainability at the Lake Ponchartrain Basin Foundation, told me that there were some thirteen different processes relating to subsidence, half of them natural, half of them human enhanced. Reducing oil and gas extraction in the area has decreased the rate of subsidence. Too, gas companies used to be allowed to discharge some of the excess fluids and brine that came out of drilling. On land, they have to put them back, reinject.

The state could lose another chunk of land the size of Delaware by 2070. To draw attention to rapidly disappearing coastline, a football field every hour, Governor Edwards declared a state of coastal crisis and emergency in April 2017. It asks President Trump and Congress to designate coastal protection projects for high-priority status and to "leverage" Deepwater Horizon oil spill money. The $20.8 billion settlement, the largest ever for the worst environmental disaster ever, included $8.8 billion for coastal restoration.

There have been other changes over the past century. In 1960, 600,000 people lived in the city of New Orleans. Population started to decline, dipped drastically during Katrina, and is now up to around 390,000. Much of the growth in the metro area has occurred on the north shore of Lake Ponchartrain in St. Tammany Parish that is not yet protected by the levee system, though there are plans to extend it there.

Before I met with Mark Davis, I took a walk along the river in Audubon Park. I had yet to see the river that defines the city, cradles it. It was low with the sand bars on the batture along the levee exposed. I watched birds flit in and out of willows. A large ship parked on the far shore. Along the park and nearby neighborhood stand stately, old houses. Long ago, the city built on the higher ground, the natural levee banks by the river. Locals refer to this area as uptown though it is in the southwestern corner. Income level moves on a sliding scale as land slopes downward, much of it below sea level. New Orleans is often described as a bowl or shallow basin between Lake Ponchartrain and a curve of the Mississippi.

In his office at Tulane, Davis, tall and trim, told me how New Orleans differed culturally from other urban centers of the South, ones over which New Orleans was once economically and commercially significant. Dallas, Houston, Atlanta were all built on growth models. But New Orleans was built on a social model, where getting along in social circles was more important than getting along in business ones. Said Davis, they "love their city but not enough to govern it, not unlike parents who love kids so much they don't want to be good parents." In a social model, change is threatening—it suggests something is wrong.

Katrina and Rita exposed the failing of that approach. And one reason it did, aside from the disastrous consequences, is that everybody had to leave. "They had to go someplace, without knowing they would come back." They went to Indianapolis, Columbus, and experienced the hospitality of those communities. It was an "astounding opening of the American heart." And then a strange thing happened. People really started to see how other cities work. Kids were now in public schools that really worked. For many, the schools and some civic institutions in New Orleans were patronage sinks.

When it came time to put things back, some politicians, including then Speaker of the House Dennis Hastert, wondered aloud if parts of New Orleans should be bulldozed.

For Davis, the issues New Orleans is facing are as much about civics are they are science. The city produced a new master plan in 2016 that was approved by a referendum of voters. It includes provisions for resiliency efforts. "I helped with the outreach effort. It was one of the most encouraging things I've seen. Once you know what you're facing, and are honest about it, you can chart a course."

But then you bump up against the issue of how to pay for it all. At one time, residents could see their way to paying for sewage systems, for sidewalks, but in the current political climate, it is very difficult to raise new revenues, which gets back to civics.

When I asked how he got into issues related to climate change and water resources, Davis shrugged, "I was a tax lawyer." Wasn't it obvious? He worked in transactional law in D.C. and Chicago but performed volunteer work for environmental groups. In New Orleans, he was volunteer counsel for the Lake Ponchartrain Basin Foundation, which deals with coastal sustainability. He then became director of the Coalition to Restore Coastal Louisiana, whose name bespeaks its mission. Along the way, he found his training in transactional law to be helpful. If there was a problem, somebody had to be working on it. If not, how do you make it somebody's job? There are steps to buying a house or business, but there were not always steps for restoring wetlands, so whose job was it to do this, and what resources are needed to make it happen?

If change does not happen fast enough, Davis foresaw a time when a crisis may force the issue, and it may be more human-made than natural. Davis told me that "long before salt water comes out of the tap, before there is seawater in your living room, the insurance industry and financial sector could pronounce judgment." New Orleans may no longer be worth the economic risk. Several panels at Res/Con addressed the insurance challenges. There may come a time when reinsurers in Europe make decisions the city has no control over. Large insurers like Swiss Re have already said they are out of the business of insuring vulnerable places. But not if you are doing something.

Mark Carney, governor of the Bank of England, told a group gathered at Lloyd's of London, a bedrock of the insurance industry, that we are out of sync with the risk horizon. "Climate change," he said, "is the tragedy of the horizon. We don't need an army of actuaries to tell us that the catastrophic impacts of climate change will be felt beyond the traditional horizons of most actors—imposing a cost on future generations that the current generation has no direct incentive to fix." He said the horizon for monetary policy extends out two to three years. For financial stability, it is a bit longer, but "once climate change becomes a defining issue for financial stability, it may already be too late."

After a traumatic event, there can a lot of assigning of blame and mistrust. Several investigations have faulted the Army Corps of Engineers for breaches in the levees that caused massive flooding. Ricky Boyett, a spokesman for the Corps, agreed to meet to discuss what happened and what they are doing now.

I met Boyett at the site of the breach at the 17th Street Canal. While I waited for him, I thought of the woman I had spoken to at lunch, still feeling vulnerable. At such times, we may need to ask for help. But what if the people who can help are also the ones responsible for the hurt? While I pondered this, a small fish briefly broke the surface, showing its dorsal fin as if a whale. Nearby, I read a plaque.

> On August 29, 2005, a federal floodwall atop a levee on the 17th Street Canal, the largest and most important drainage canal for the city, gave way here causing flooding that killed hundreds. This breach was one of fifty ruptures in the federal Flood Protection System that occurred that day. In 2008, the U.S. District Court placed responsibility for this floodwall's collapse squarely on the US Army Corps of Engineers; however, the agency is protected from financial liability in the Flood Control Act of 1928.

Boyett, in his thirties and wearing sunglasses, met me in a gravel parking lot of Station 6, a restaurant named for the pump station. Walking across the bridge, he pointed out the site of the levee break. New concrete could be seen next to old. Boyett said the flood was the result of a "design flaw." Rather than water overtopping the levee, it eroded from underneath. Water was rushing out of the canal from all the collected rain, headed toward Lake Ponchartrain. At the same time, water was rushing in from the storm surge and lake, creating a massive, swirling eddy system that eventually found the weakness in the soil under the wall.

On the other side, I could count a dozen Caterpillar pumps. They had large black tubes like closed water slides. After Katrina, Boyett explained, the Corps built interim closure structures at outfall canals, closing off flow from the lake in case of surge. Of the temporary pumps, a few had fraying tarps over the engines, protecting them from rain. They test the pumps once a month, twice during hurricane season. Sometimes, fish like the one I saw are pulled out through the pumps. When pelicans hear the diesels rumble, they gather as if responding to the dinner bell.

Beyond the temporary structure was a larger brick structure with seventeen pumps. These can handle twelve thousand cubic feet per second (cfs) and are designed to handle a one-hundred-year storm. He said they use models of sea level rise and that standard should hold through 2057. The cost of this and two other structures was $650 million. The temporary structures were $400 million. The Corps partnered with the state, which matched 35 percent. The Sewage and Water board took ownership ahead of the 2017 hurricane season.

Boyett thought of it being not foolproof but part of hurricane risk reduction. "It's not a life-saving system. It is for defending an empty city in a storm."

After I talked with Boyett, I drove to the crown jewel of the levee system, the $1.1 billion Lake Borgne Surge Barrier, twenty-six feet tall and nearly two miles wide. Nicknamed the Great Wall of St. Bernard Parish, it is supposed to protect the city's eastern flank from Lake Borgne if the water rises. It should prevent what happened during a Katrina, when a fifteen-foot-high surge of water tore through canals and toppled levees, swamping the Lower Ninth Ward.

To get there, I drove through eastern New Orleans. Many of the houses there had been hit badly, windows broken and roofs torn off. The devastation was so bad I wondered if some could still be from Katrina, but people were walking around and stores were open. I learned from one that a recent tornado caused the damage.

I drove down through an industrial area to where my phone's map said the surge barrier was, down Intracoastal Waterway Drive. I passed several large industrial facilities, shipping hubs, and kept going. It appeared that I might have been able to drive right onto the gate though a sign said "authorized personnel only." I wanted to see what a billion dollars looked like but I didn't want to go to jail, so I walked to the base of the levee. It was just me and a survey truck from the Southeast Louisiana Flood Protection Authority.

Maybe he could authorize me. But as I approached the truck, I sensed I was disturbing some sensitive information gathering, or napping, so I kept walking to the top of the levee. A billion dollars looks like a lot of concrete. And a yellow gate to let ships through. And a lot of engineering I could not see.

High on the berm, the wind picked up and I had a good view of the ship canal and the wetlands. The barrier acts as the wetlands once did. Vegetation is not a perfect barrier, but it slows the water down, provides some resistance.

After walking on for a while, I drove over to the Bayou Savage National Wildlife Refuge. The refuge is within the city limits and flood protection system. It is a flashback to what once was, but because it is within the system, the natural flow of water is disrupted. To remedy this, networks of pumps and flap gates recirculate the water, mimicking the seasonal water levels. And each year the city of New Orleans airlifts recycled Christmas trees into the refuge to act as wave breakers in open ponds, allowing marsh grasses to grow. I had a canoe with me and I wanted to paddle in the bayou, but the wind was really strong. Those slow-moving bayous look inviting but also are easy to get lost in.

Much of the bottomland forest, up to 90 percent of the trees, was destroyed by Katrina—more by the saline water than the intense winds. What I was looking at was a landscape in recovery. But a resilient one.

As those efforts continue, the surge barrier will help prevent future destruction, but paying for upkeep is an issue. The Corps has agreed to pay for some, but most of the cost of maintaining falls to local governments. But in St. Bernard Parish, voters have twice voted against tax increases for hurricane protection, of even five dollars a month—a six pack, a pound of crawfish, back to civics. A new bill would allow the water authorities to tax properties without voter approval.

Still, while voters did not want to pay for their share of the levee, others in New Orleans were proud of their $14.5 billion levee system. Gen. Russell Honoré told me at a lunch meeting at the conference that the city had a story to tell, that they hired the "best in the world." Honoré coordinated military relief efforts during Katrina. He took charge when civilian leadership failed. In video clips, he can be seen asking soldiers to put away their weapons. And what Anderson Cooper called looting, Honoré, a Creole from Point Coupee Parish, called survival. Some nineteen thousand people gathered outside the very convention center we were in, huddled under the overhang. Some of the exhibits were a reminder of that time: Aqua Fence (a portable flood barrier), FORTS (portable shelter systems), self-heating meals, and something called Flo-paq, a flexible container for liquids.

I stayed for some afternoon sessions, but it wasn't long before I was ready to get back outside, explore. Being in the convention center reminded me a little of shopping in the way it tired me. Retail presenters put their best face on what they are selling, but do you get the unvarnished truth? So we fueled up on coffee and beignets from Café du Monde and hit the road. I wanted to see the Isle de Jean Charles because the governor talked about it and because I had read about the island and its people as being among the first climate change refugees.

We had good expertise for our task. With me were three students in their first year of college: Dean, a geography major, Cameron, interested in architecture but also sociology, and my son Sam, a wildlife conservationist. Species would be threatened and environmental justice and planning were at the heart of the issues of displacement as communities disappeared from the map. As the English professor, I promised to throw in a visit to Grand Isle, the setting for Kate Chopin's *The Awakening*, if time permitted and we didn't get lost.

On the way out of town, we passed a graveyard and above-ground tombs. In *Life on the Mississippi*, Mark Twain observed that the people of New Orleans could not have cellars or graves, "the town being built on 'made' ground; so they do without both, and few of the living complain, and not any of the others." The others do not complain, but in floods, caskets slip from their crypts, and they bob and float down the watery streets.

We followed the river west out of town. The area we traveled through was known as "Cancer Alley" because over one hundred petrochemical facilities and refineries are interspersed with poor, historic residential settlements with higher than average rates of cancers, miscarriages, learning disabilities, and other ailments.

If we were college students studying the land use patterns, we would learn that the landscape along the river began to change in the 1700s with French settlers, transforming bayous, meanders, and floodplains into geometrical strands or arpent long lots, perpendicular to the winding river. After the Civil War, Twain's time, freed slaves were granted land nearby sugar, indigo, and cotton plantations. Freedmen's towns began to form. One hundred years later, many river plantations were replaced with petrochemical plants, though some of the residents remained in the shadow of chemical processing.

In *Petrochemical America* (2012), photographer Richard Misrach and architecture professor Kate Orff explore the social, environmental, and health impacts of the petrochemical industry in Cancer Alley through photography, writing, and map illustrations. One map overlays toxic release sites in expanding circles to show a million pounds of release beside human figures to show population density. Another shows the chemicals produced, the propylenes, ethylenes, acids, rubbers, and polymers at exact locations along the river. A map of chemical names and their symbols.

Oil companies have routed an estimated twenty-six thousand miles of pipeline throughout the southern coastal wetlands. A web of canals has been cut through marshland to bury the lines. These connect to offshore drilling platforms and to refineries and chemical processors. The network of pipelines crisscross vast spans of land and water, realigning the needs of ecologies, communities, and commerce into a new geography. The gulf's salty tides surge into the pipeline arteries. Instead of clogging, they widen, erode.

Gradually, the sprawl dissipated, and we traveled on a small finger of land, water on both sides. On a state map, we were in the boot part of the Louisiana, though the map on state road signs, the iconic image of the state, presents the area in solid form. Around us there seemed little terra firma. We entered an area beginning to look less like a boot and more like a frayed high-top canvas sneaker, ragged at the edges with worn tread. Zero-in on these places on Google Earth and whole communities, such as the one Bob Dylan sang about in "Tangled Up in Blue," "right outside of Delacroix," seem to float atop the marsh grasses. NOAA has released updated nautical charts that removed forty place names in a single county, Plaquemines Parish, just south of New Orleans.

We used the GPS and digital mapping tools built into our phones to navigate. The road took us toward Houma, which the phone voice pronounced in a way that could either be taken as offensive, *homo*, or as the genus we four in the car share. The problem with phone maps, in addition to pronunciation, is that they give a kind a tunnel vision, where you are and where you are going, with little left for the imagination to roam, connections to be made between one place and another, a wider perspective.

We kept driving until we came to the Intracoastal Waterway. The draw bridge was up and the road was closed so we had to turn around. While we waited for phones to reroute (though it kept suggesting a U-turn), we asked for directions. We had a canoe on top of the car for the spring breakers to take into the bayou for a few days to camp. A man out retrieving his mail asked where we were going to fish. The canoe was our calling card.

General Honoré had joked that people out on the fringes of the coast used to see alligators. Now they see sharks. Ask some of them about climate change, he said, and they are likely to resist. But ask them how the fishing has changed in the past twenty years and prepare for an earful. The man gave us directions, and fishing advice, in a dialect pleasing to us all, though we were not sure we understood any of it other than go back to the stop light in town and turn left.

By now I wanted a map. Not phone directions but an actual map, the original app, which has been in use at least since *homo sapiens* painted in caves. The problem is they are not easy to find. I checked at gas stations and found plenty of cell phone chargers and accessories, lots of pork cracklings, extra spicy Cajun cracklings, bags of rubber worms and bait, but no folds of paper delineating geographical boundaries to be used in wayfinding.

We found our way there using what little charge we had left. The spring breakers showed me a feature of the iPhone I did not know existed. Press on the compass app and you can also get elevation, though at one time Siri told us that we were thirty feet below sea level. It was possible, but we were looking at the sea on either side, and it did not feel as if we were underwater.

Brett Anderson, a writer for the *New Orleans Times-Picayune*, wanted to update maps to reflect the loss. His "Louisiana Loses Its Boot" investigation led to an alternative map. He had Andrea Galinski, a coastal resources scientist at the Coastal Restoration and Protection Authority, the entity overseeing the master plan, visualize what Louisiana would look like if only "walkable" ground appeared. In the resulting map, "the boot appears as if it came out on the wrong side of a battle with a lawnmower's blades." It loses a chunk of the heel, a gash in the arch, and there are a few scrawny toes where the delta would be. Ponchartrain and Borgne flow continuously into the gulf. Still, there are firm edges that cannot really be represented as lines. In a follow-up to the original article, Anderson spoke with Jeff Carney, director of LSU's Coastal Sustainability Studio, who said the lines at the bottom edge are neither land nor water but wetlands, and they should be represented as more like a "zone that will continue to change as we try to reverse the damage."

Carney, who thinks a lot about how to communicate coastal vulnerabilities, told me that for "a hundred years we've trashed the marsh, ignored it. Now we value it, so how do you communicate that value?"

Anderson has acknowledged the map's flaws. "My intention was to argue that maps are powerful tools to get people understanding an important subject," he said, in a follow-up. But could a map that eliminates wetlands communicate the wrong impression, as healthy wetlands are joined with ones that are being lost to erosion, subsidence, and sea level rise? What might help are maps that show where populated areas are subsiding, or coastal lands at risk but that might still be saved, or a map that helps people more accurately assess flood risk from sea level rise or storms.

The Isle de Jean Charles is located on a vulnerable spit of coastal land southeast of Houma. It is home to Native Americans, about twenty-five families, a mixed tribe of Biloxi-Chitimacha-Choctaw. They were originally forced to flee *to* the island as refugees from the Indian Removal Act in the 1830s. Now, they may be asked to leave because of climate change.

We crossed a narrow causeway. There were signs telling us not to drop anchor because there were pipelines nearby. One homemade sign hanging diagonally off a bamboo pole said "Private lease. No crabing [*sic*] at anytime." And this one greeted us as we entered the island, "We are not moving off this island. If people want to move they can. But people have a right to live where they want not where people tell them to go." It continued in all caps, ink fading. "They say the island is fading away soon . . . If the island is not good stay away."

The creator of the sign was returning from fishing, riding his bicycle while holding his rod beside him. We debated talking to him. In the rural area where we live, I wouldn't normally just intrude on folks, nor would they like to be intruded on as they are suspicious of strangers. And the sign gave a kind of warning, stay away. But only if the island is not "good," and it was good to us. I wanted to jump right into the issues. What do you think about leaving? Where will you go? But I knew the question that usually got fisherman talking, "doing any good?"

Edison Dardar, late sixties, frothy white hair like sea foam, took a look at my car, canoe on top. "Only a few mullet." He headed up to his house now, hosing off boots and gear.

"What do you normally catch?" I approached slowly.

"Drum, flounder, trout, redfish. Lot of good fishing there," he said, pointing to area behind his house, as if inviting me to try my luck.

We talked some more. I could sense the pride in place, the land going back generations, even though some 98 percent of it has disappeared. Only a few structures remained, most on stilts. For many, you had to cross rickety bridges over canals to get to them. I finally asked about relocation.

"Why would I leave? There is fish here. People have always fished here, many in a pirogue," he said, pointing to the upside-down contraption on my car's roof. "They want me to go to Thibodaux. What am I gonna eat there, chicken?" He said *chicken* derisively.

I asked about the money for relocation, a lot of it.

He offered mistrust, "They ain't gonna give us no money."

The island is thought to have been named after the father of a Frenchman who married into the tribe in the 1800s. The island road was not built until the 1950s. Access was only by water, so many continued to speak a French-Cajun dialect. Some of the land loss was evident through the dying trees. Albert Naquin, chief of the Isle de Jean Charles band, told Brett Anderson that the reason they live there is that "we're forced to. We're frickin' Indians."

I told the spring break crew about the reasons for the land disappearing: erosion, rising sea levels, lack of soil renewal, subsidence, and changes due to dredging for oil and gas pipelines. At the edge of town was an orange escape pod, shaped like a UFO, that you see on drill rigs and ships, to be used in case of a flood. The road out frequently floods during storms.

The people on the island have been touted as the "first climate change refugees," but taking in all those factors, we wondered if it was accurate. General Honoré said much of the loss is due to industrial damage. Are they industrial damage refugees? The Corps has said putting a levee around them is not cost-effective. But why? Because they do not produce enough economically? Are they then not political or economic refugees, because they did not have the clout to push for flood protection?

In 2013, the Board of Commissioners of the Southeast Louisiana Flood Protection Authority filed a historic lawsuit against more than one hundred oil, gas, and pipeline companies, demanding that "the catastrophic effects of the oil and gas industry's canal dredging be abated and reversed and the damage to the coastal landscape be undone," seeking billions in damages. In 2015, U.S. District Judge Brown ruled the authority's levees were too far from, or indirectly affected by, the damage allegedly caused by the industry. She also said the authority had no right to sue under permits issued by the state or the U.S. Army Corps of Engineers that allowed the companies' energy exploration or transportation activities in the first place. At the time of the suit, then governor Bobby Jindal said it would damage one of Louisiana's major employers. About half the people I spoke with in Louisiana agreed with Jindal. Others said that since the oil companies contributed to the land loss problems, and profited from the activities, that they ought to be held accountable.

In *Strangers in Their Own Land* (2016), sociologist Arlie Hochschild traveled to Louisiana to try to understand the "red state paradox." Louisiana is among the states near the bottom of ranks in human health and education statistics, yet they vote for candidates who reject government intervention. Red states tend to be poorer and rely more on federal help, but they are also more opposed to the federal government.

Hochschild finds that such voters feel their cultural beliefs are denigrated by the culture at large. They feel that they are seen as rednecks and that they live in a region that is also discredited. Many of them are deeply devout, but they see the culture at large becoming more secular. And then they see an economic trapdoor that used to swallow only minorities but is now opening for them. Altogether, they feel like "strangers in their own land." And they feel the government has been an instrument of their marginalization.

The people Hochschild spoke with arrived at that realization through three key routes: faith, taxes, and honor. They felt the government curtailed the church, it took too much in the way of money, giving to those who did not deserve it, and it took something like their pride. But they have no language for victimization, believing in a kind of stoicism and hard work that the government removes. As for climate change, many either did not believe or thought it a "Trojan horse" that would further expand government. Many who worked for petrochemical companies also hunted or fished and knew the waterways were polluted. But they didn't get too hung up on a nostalgia for a cleaner time. Besides, many had an eye on the long-term, an eternity even. "We're on this earth for a limited amount of time," one resident told her. "Heaven is for eternity."

When President Obama put a moratorium on drilling after the BP spill, many became madder at the moratorium than the spill. "The spill makes us sad, but the moratorium makes us mad," one woman said. As much as they hated the government, they loved the private sector more. In some cases, the greater the risk to environmental pollution, the more likely people thought that the government was overreacting to the issue. Writes Hochschild, "The Louisiana story is an extreme example of the politics-and-environment paradox seen across the nation." As bad as environmental pollution is, cultural pollution is worse in the eyes of many she spoke with.

The "deep story" she uncovers is one of a perception of line cutting. People imagine they are standing in line, playing by the rules, waiting for their turn at the American dream, but others are cutting in line. And some of them are minorities, a different race or gender, or from other countries. And some are birds. The brown pelican, the Louisiana state bird, was once nearly wiped out by pollution but in 2009 it was removed from the endangered species list, a year before the BP spill. To survive, it needs protection, clean water, air, and oil-free marshes, ones safe from erosion. You need food and water too, but "it's just an animal, and you're a human being." You no longer feel pride in a government who rewards these cheaters, so you band with others who also feel estranged.

John Lopez, of the Lake Ponchartrain Basin Foundation, said there have been some studies saying oil and gas are responsible for 30–45 percent of the land loss. But in Terrebonne Parish (where the Isle de Jean Charles is), he has read that it is as much as 80–90 percent responsible. Before he became interested in saving coastal wetlands, Lopez worked as a geologist for "one of the bad guys." He worried that, as many of the big wells have been tapped, the bigger companies retreat and the smaller operators move in. Already, many live in Houston but fly in to do work. They do not have as much "skin in the game," and if spills happen, or canals are dredged inappropriately, they may not be able to pay.

Whether oil and gas companies pay or taxpayer-funded Department of Housing and Urban Development (HUD) pays, Mark Davis, of Tulane, said the issues facing the residents are complex, not written off with a check. Who is in? Who is out? Who owns the lands they leave behind? What rights do they give up, including mineral rights? Which do they keep? "It's not a simple matter of moving noble, hard-pressed people to another place. They have to be willing to go. The neighbors have to be willing to accept them. There has to be something for them to do." And if you don't know the answers for a couple of dozen people, it doesn't get any easier for New Orleans or Miami.

We drove on through moss-draped oaks and shoulder-high marsh grasses, looking for someone else to talk to. Night was falling, and two great horned owls appeared in bare-limbed live oaks, likely drowned by salt water. We got out of the car slowly, closing the door quietly, shushing the beeper that lets you know your keys are still inside, hoping to hear a call or snap a picture. But the owls flew away, as if having too much dignity to be photographed.

We marveled at the silent flight. That they have been painted in caves, haunters of burial grounds, omens of death and destruction is likely because they are creatures of night, and therefore misunderstood. Owls can punch above their weight, taking species larger than themselves. They are certainly not *chicken*.

Leaving the island, we found an open door, lights on, someone to talk to. Two other Dardar kin, brothers of Edison, sat in chairs in a modest shack, drinking beer and watching the night fall. Wearing a Saints cap, one said friends call him "Turtle." A bolt action rifle lay on the table in front of them.

This time, I used the owl sighting as opening gambit, asking them if they saw the pair. The brothers assured me they were gentle and would not hurt us. The spring breakers joined me, all of us crowded in their doorway. A tiny green frog jumped on the top of a chair. We all noticed. I brought up relocation, the money to move.

"We're not leaving. Nobody signed papers," said one.

"Why leave?" said the other. They had cold beer and a freezer for their bait. The radio played "Bennie and the Jets." Both of them smiled, wide beatific grins, not saying much, but pleased with their shack and beer and bait at the fag end of the world. Disappearing or not, rich or not (not), those brothers were happy, and it was a sight I am glad the college students experienced. Not unlike those owls, they were content in the night, and wanted to be left alone.

The next morning, I dropped the spring breakers off to canoe and camp among a plexus of bayous. After making sure they had all necessary provisions and safety equipment, I drove on to Texas, crossing the state line near Beaumont. The oil strike at nearby Spindletop in 1901 transformed this once rural state. I headed on to Houston, beneficiary of the boom, now the largest city in the southern United States.

In the morning, I met with John Anderson in his Rice University office. Houston is a busy place with lots of traffic. But the pace on the Rice University campus and in his office was slow, casual. Though Anderson has bundled into parkas to study the cold of Antarctica for some forty years, today he wore a tropical shirt. He first noticed melting trends when studying in graduate school, but they did not call it global warming or climate change then. He talked about the difficulty of measuring loss in the Antarctic. In Greenland, you can measure glaciers in relation to land. But in Antarctica, they use sophisticated equipment that measures thickness of ice, altimetry. What that shows is that western Antarctica is losing ice at a high rate. He said that while the eastern part is stable, "not one informed glaciologist in the world would say west Antarctica isn't losing significant mass, experiencing thickness variations." The great fear is what is happening underneath from warm currents, where the ice is minus eight Celsius, but the water is half a degree above freezing.

For a scientist who deals with models and projections, he emphasized the unpredictability of the Antarctic. How do you put a number on what could happen? "It could collapse within a decade. We don't know." He talked of current efforts to look at the Thwaites glacier. And he referred to a study he had read recently in *Nature* by Robert DeConto of the University of Massachusetts Amherst and David Pollard of Penn State. Looking at temperatures, carbon dioxide levels, the loss of ice cliffs off the glacier, they created a model to reproduce high sea levels of the past, such as a climatic period about 125,000 years ago when the seas rose to levels 20 to 30 feet higher than today, calibrating climate and data sea level estimates and then applying them to future greenhouse gas emission scenarios. They concluded that "Antarctica has the potential to contribute more than a meter of sea level rise by 2100 and more than 15 meters by 2500." Anderson said seas could rise by as much as 60 meters if the whole ice sheet goes. "We're not used to being alarmist in the scientific community. But we're seeing changes over the last few decades that are measurable geologically."

Anderson also researches coastal systems, and 85 percent of the Texas coast is barrier islands. It's a low-gradient coast, and Anderson has researched how barrier islands, past and present, have responded to inundation, which is also difficult to measure. Anderson rattled off about a dozen different variables that influence how islands respond, including width of the island, offshore gradient, height, size of the sand particles.

The Texas Bureau of Environmental Geology has been taking aerial photographs longer than any other state. Edgar Tobin was a World War I flying ace. He took over an aerial mapping firm in 1928, helping to survey the state to develop the oil industry. Some of the aerial photographs were shot out of a biplane with a box camera. While one brother flew the plane, the other hung out over the wing to capture the ground below.

These images can be used to track shoreline retreat. One island, Follett's, just west of Galveston, is retreating at a rate of four meters per year. The rate of shoreline retreat has doubled. If you double it again (as sea levels rise), the island is gone by the end of the century. The unknown is storms and their intensity, but scientists have a rough idea of what can happen, including overwash, and models Anderson has run suggest Follett's has about two hundred years left. Anderson and other geologists know that the island is two thousand years old. If it was retreating all along at the rate it is now, it would be four kilometers inland. "Barrier islands don't wait to be submerged. They respond radically."

Anderson referenced the work of Benjamin Horton, a professor in the Department of Marine and Coastal Science at Rutgers University. Looking at fossil data in coastal marshes, including plants sensitive to salt, Horton has researched sea level records going back two to three thousand years.

That data shows a long-term rate of a half millimeter per year of rise. But then it ticks up abruptly in the late nineteenth and twentieth centuries, to where the rate increases five- or sixfold. Anderson said he gives talks at local civic groups. When he mentions three millimeters of rise per year, someone will hold up their finger and thumb, showing that amount. "That's this much. It'll take forever before Galveston is underwater. We'll fix the problem by then." But when you put that in the context of geology, the way that rate has changed, "that's the most compelling evidence for climate change." You can only change the rate by melting glaciers and warming oceans.

Thinking about all the challenges, our conversation turned to politics. I had heard two things on the radio that morning: 1) in spite of all the challenges Texas faces in terms of drought, wild fires, and flood, they were considering a bathroom bill similar to the one passed by North Carolina, and 2) Texas leads the nation in wind production.

In the wind story, Rick Perry, then governor and now secretary of Energy, was given credit for signing a bill to lay the infrastructure, having citizens pay for it in energy bills, in effect, socializing the costs. The need for alternative forms of income was prompted by droughts. Ranchers could not produce enough feed to keep their cattle alive. Now, landowners are making as much as $10,000 a year per turbine. The heck with the cows. Climate change, however, is not part of the discussion on converting to wind. Former mayor of Sweetwater, Greg Wortham, said, "That would defeat the purpose. If you walk around saying I'm green, I'm green, you lost. You create divisions."

Anderson told me the conversation about climate change was "almost forbidden," and he told me a story. One day, he was sitting in the very office where we were chatting and the president of the university, David Leebron, showed up at his door. "You're being sued." Anderson never got a summons, but "talk about your day going to hell in a basket." The story came out in the *Los Angeles Times*. He was being sued for "breach of contract," though he had no contract.

The Texas Commission on Environmental Quality (TCEQ) censored him. He had written a chapter in a book as outreach. That chapter was sent to the TCEQ for review. They struck out references to the Intergovernmental Panel on Climate Change (IPCC), global climate change, and a figure that showed sea level rise. "After that, there wasn't much left." The president said he was proud of him and that the entire legal staff would back him. The legal team found an agreement but no real contract, and when Rice threatened to sue for defamation of character, TCEQ dropped the suit.

Anderson has been part of a group convened by Sen. Sheldon Whitehouse of Rhode Island to talk with state officials about the science of climate change. He has gone with Katherine Hayhoe, an atmospheric scientist and climate change communicator from Texas Tech. Twice, Anderson has traveled to Austin on these errands. "No one came. Not one legislator." What bothers Anderson is the accusation, like the one about the book chapter, that he is in on some conspiracy. "Science is the counter of conspiracy. We disagree to make headway. The conspiracy is not among the science community. It's among the people who are denying it for other reasons, because the scientific documentation is indisputable." Just across the state line in Louisiana, Anderson has a friend who teaches. The kids in school there know. But when Anderson speaks to local Texas schools, children have no prior knowledge. And yet, the major oil and gas companies say it's real, that it's something we have to deal with, and it's caused by humans. Visit the websites of Shell, Exxon, Chevron, and they all have statements on reducing climate change risks. Of course, as Sheldon Whitehouse has written in his book *Captured: The Corporate Infiltration of American Democracy* (2017), these companies can advocate publicly for positions in the public interest, but the trade organizations representing them may do the opposite. They get to look like they care but are also making sure nothing is adversely affecting business interests.

After speaking with Anderson, I walked across campus to find Jim Blackburn striding into his office after a class. In addition to his name, there was a lot about him that was black, including his jeans, shirt, cowboy boots, his mustache, but he was also full of optimism and good humor about the Texas coast.

I wanted to talk to Jim because I had read about a continuing education course he was offering at Rice University, "Full World: Houston's Economic and Ecologic Future." Jim is aware of his audience. People who take the course are often retirees who have had long careers, perhaps conservative in the way they would rather things not change too quickly, though the future, from Blackburn's perspective, "will be nothing but change."

Jim's course talks about how rather than have an "economy expanding to the moon," which Houston's has done, as has its development patterns, "we have one aligned with natural cycles." And yet, he hopes that businesses and corporations will make that change.

The big idea Blackburn promotes is carbon sequestration by ranchers. He wants to restore native prairie, build up beautiful organic soil. In healthy black soil, that black is the carbon. And he thinks oil and gas companies will one day pay ranchers for carbon rights.

I asked how that would happen, and he said the businesses themselves might do it. Consumers might force them to do it. In ten years, "we'll be on our way to a carbon economy. Carbon in the ground will be a commodity." He could see the entire agriculture industry restructuring around this. Fuels come from pressed organic matter. He wanted to press that carbon back into the soil.

He said he found his optimism in the business community, "in their pursuit of money." They will figure out what will cost them money, and if they can make money. "Money is where the green is," meaning money and ecology can become equated. He thought the oil and gas industry would come to the conclusion that carbon is "strategic," just as Coke has with water. Facing shortages and accusations of exploiting freshwater supplies, including protests, Coke got strategic about their access to water.

Blackburn said the carbon will be stored not in deep formations, where it would cost $100 dollars a ton. He thought it could be stored in soil for $40 a ton. Farmers and ranchers would plow up grain and plant prairie seeds. Using a carbon monitor, they would measure before and test again a few years later. "Sell the difference in carbon stored." Carbon absorbs into plants by photosynthesis, then into deep black soil, a natural sequestration. You can measure grams per cubic centimeter and extract from there, but Blackburn thinks they can store two to three tons per acre per year. That could mean a hundred bucks an acre per year in income. "Every farmer in Texas would do backflips for that." If you keep cattle on the land, you net out methane emissions.

Another Texan, Secretary of State James Baker, namesake of Rice's Baker Institute, has called for a carbon tax. It would replace the Clean Power Plan, raising prices of carbon to bring down consumption. While not acknowledging the science of climate change, Baker recognizes the risks of it, and he told the *Washington Post* that his plan is "simple, it's conservative, it's free market, it's limited government."

I asked Blackburn how he got interested in carbon sequestration, and he gave me an unlikely answer: hurricane surge flooding. Blackburn wrote a book, *The Book of Texas Bays* (2015). He's an avid birder and proud of the twenty-four million neotropical birds that migrate through the double canopy forest in the coastal areas. He targeted two million acres he did not want developed, below the twenty-foot contour line. But he knew it was unlikely for that area to be regulated in some way, preserved, so he started looking at other means.

But Blackburn acknowledged that market solutions will not solve everything. There are neighborhoods in Houston built close to bayous that flood every other year. He told me some statistics he had gathered for his class, that in the last thirty years, they've had had thirteen one-hundred-year storms, five five-hundred-year rainfalls. The building and planning codes are based on statistics that are obsolete because of climate change. He said 25 percent of Harris County is in the floodplain. An honest map puts half the county in it. "We don't want to deal with things. It's easier not to deal with it, deny, put it off."

I asked him about denial, and he said some of it was tied to the community he put his optimism in. "For some people, if we admit to climate change, we're adversarial to the oil and gas industry." He said Houston was a transient place, one people did not have long attachments to. The oil companies prefer it that way, he said. Then again, environmentalists have not been good at going in the direction of thinking about money. There's a Puritanical streak that runs through the environmental community, "where money is sin."

One difference between Texas and Louisiana, aside from the land loss, which created an urgency, was the existence of a commercial fishery in Louisiana. They kept their eye on things, noticed changes. That's why Blackburn felt that they had a plan and Texas did not. "We have no plan, no map for what is to come. I have plan envy," he told me, which is how he begins his new book, *Texas Plan for the Texas Coast*.

Part of that plan came back to business. The way to keep farmers and ranchers in farming and ranching was in carbon trading. That would solve other problems, such as the dead zone in the gulf, because there would not be a need for fertilizers. And those natural lands would provide a storm buffer and would help alleviate flooding because they were not developed.

I began that morning with a tour of the Houston Ship Channel on the good Sam Houston. The channel has been widened and deepened by dredging Buffalo Bayou and Galveston Bay. It's one of the busiest seaports in the nation. Petrochemicals are moved along with midwestern grain. Passengers embark at Turning Basin in eastern Houston, the head of the channel and most upstream point to which general cargo ships can travel.

We slipped from our mooring and motored into Buffalo Bayou. Through brown water filled with flood trash, we sailed by mountains of metal, twisted, rusty rebar to be used in recycling. Huge cranes scooped scraps of metal from one pile and shifted them to another, spitting out clouds of rust. On shore metal crunched, jackhammers drilled, and trains squealed. Above the noise and the thrum of the engine, I could hear the voice over the loudspeaker giving out the facts, such as which tanks belonged to whom and that the Port of Houston was the busiest in the nation in terms of foreign tonnage. There were over a hundred storage tanks and large ships from Hong Kong and Panama, whose names made a kind of ironic word art—GLOBAL HORIZON, OCEAN BEAUTY.

Gas flares blazed into the cloudy sky. Between the sinister spires peeked the downtown skyline.

One of the slides Blackburn showed me was about hurricane intensity. With the warming ocean and increased moisture, hurricanes could become more intense. One of them could hit the heart of the American petrochemical industry. In an op-ed for the *New York Times*, writer Roy Scranton imagined a storm crashing through refineries, "cleaving pipelines from their moorings, lifting and breaking storage tanks," releasing a flood of jet fuel, sour crude, and natural gas into parks, schools, offices, neighborhoods. Blackburn said a storm like that would "burp out the nastiest stuff you've ever seen."

Two main research teams have led the way in preparing for such a storm. Blackburn and Rice University's SSPEED Center (Severe Storm Prediction, Education and Evacuation from Disasters) have advocated for a layered defense, including a mid–Galveston Bay gate that could be closed during a storm and one midway up the channel. SSPEED's models show it could stop a surge at a cost of a few billion dollars. Blackburn worried that a larger system could impact the bay as big structural interventions run the risk of a damaged ecosystem. But he said, "I have arrows all over my body" for supporting a gate in the bay that didn't directly protect houses.

The day I arrived to meet him, the *Galveston Daily News* reported that there was consensus on another plan. George P. Bush (eldest son of Jeb), Texas land commissioner, said in a speech to the American Shore and Beach Preservation Association that he endorsed a plan out of the University of Texas, Galveston, dubbed the Ike Dike. That plan is the brainchild of Dr. Bill Merrell and calls for a fifty-five-mile-long "coastal spine" along the gulf.

I drove down to see Merrell the next day. I-45 ends in downtown Galveston, once known as the "Wall Street of the South." The storm of 1900 wiped out much of the town, and Houston, fifty miles north, took over as the center of commerce. Population shifted north, though by the looks of the development along the corridor, it was again extending south. Big box stores and fast-food franchises gave way to beach bars and historic homes, kitsch and southern elegance rolled in one. The gulf spread murky on the horizon.

Merrell is the George P. Mitchell Chair in marine sciences at Texas A&M. After making billions in pioneering fracking to free natural gas from shale, Mitchell, a geologist born to Greek immigrants, moved to Galveston, turned his attention to philanthropic activities, including historic preservation. He donated the 135-acre campus on Pelican Island to Texas A&M. The causeway into Galveston over the bay is named for him and his wife, Cynthia Woods Mitchell. Mitchell once sent Anderson a letter congratulating him on his 2008 book, *Formation and Future of the Upper Texas Coast*. In his later years, Mitchell became an advocate for coastal resiliency. In 2001, he changed the mission of the Houston Advanced Research Center to focus on sustainable development. There were similarities between his story and the Rockefeller Foundation.

On the campus at Texas A&M University at Galveston, a school focusing on maritime study, cadets, affectionately known as "Sea Aggies," walked around in khaki uniforms, similar to those worn by naval midshipmen.

I met Merrell in his office overlooking a small harbor. He was surprised but pleased to learn of Bush's support. It had been almost a ten-year process. Merrell proposed the idea not long after Hurricane Ike struck in 2008.

During the storm, he was in his office on the second floor. There was eight feet of water in his building, fourteen feet going down the Strand, the name given to the historic district. *Strand* comes from an Old English word for *shore*, and means *beach* in German and Dutch. The street, also known as Avenue B, runs parallel to Galveston Bay.

A restaurant in the historical district posts the high-water mark for the 2008 Ike above my head, then the hurricanes of 1900 and 1915 at eye level.

Merrell's coastal barrier is called the Ike Dike because it would forestall the devastation that happened as a result of Hurricane Ike (2008) and ones that preceded it. It would be similar to one the Dutch have used to protect the Netherlands, and Merrell has been in close consultation with Dutch engineers in the conceptual design of the structure. They were arriving to meet him the following week.

The proposed barrier would extend an existing wall. After the hurricane of 1900, Galveston built a seawall seventeen feet high (the storm surge was sixteen), and they raised buildings out toward the coast by as much as thirteen feet. Everything else sloped down and away from the wall and shore. The cost was $6 million, about $150 million today.

But it worked. A hurricane came in 1915, "as if a test of the resilience" of the citizens of Galveston, said the voiceover of *The Great Storm* at Pier 21. The storm claimed eight people, but the city stood intact, "in defiance of the power of the hurricane."

In 1900, a third of the population died, about ten thousand people, still the nation's deadliest natural disaster, even post-Katrina. At the close of the nineteenth century, the city's natural harbor made it the center of trade in Texas, rivaling New Orleans as a cotton port.

At the time of the storm, there were no good warnings. The Weather Bureau methods were still evolving, and forecasters had no way to know where the storm would hit or how intense, and they did not like to panic residents who may or may not be in the storm's path. In *Isaac's Storm*, author Erik Larsen credits Isaac Cline, the island's meteorologist, with violating Weather Bureau policy and issuing a hurricane warning. The warning came too late, however, to allow residents to evacuate the island. The barometric pressure was the lowest ever recorded. Winds were measured at one hundred miles per hour, but the instrument for measuring it was blown off the building. Fierce winds blew out glass and slate tiles from roofs, which became missiles. For some, the safest place was in the stormy water so as not be impaled by flying debris.

After the storm subsided, bodies were carried out to sea on barges, weights attached, and left to sink to the bottom, only some came back to shore. Ida Smith Austin, a Sunday school teacher living at the prominent Austin house on Market Street, gave an account of surviving the awful night.

> The mournful dirges of the breakers which lashed the beach, the sobbing waves and sighing winds, God's great funeral choir, said their sad requiem around the dead. The sea as though it could never be satisfied with its gruesome work washed these bodies back upon the shore, the waves being the hearses that carried them in to be buried under the sand. The terrible odor from the thousands of putrefying bodies was almost unbearable.

The corpses had to be burned. Huge pyres were made, scattered lumber and wreckage thrown on top, townspeople held at gunpoint unless they participated in the stacking of cadavers before they were incinerated.

The wreckage of not only Ike but the 1900 storm reaches deep into the cultural memory. For Merrell, one reason to build the Ike Dike was that he saw firsthand how the storm disrupted all public services, causing a great social upheaval. Public housing had to be bulldozed because mildew was so bad. The poor and elderly were badly affected. Structures like the Ike Dike may not be able to stop any and all storms, but in Merrell's view, "they take you from a catastrophe to a disaster, a total wipeout to at least some functioning services, social structure, police, firemen." People do not have to die because they cannot get insulin or to the hospital.

For Merrell, complete recovery is not possible, so he is focused on prevention. He feels like New Orleans and then Sandy exposed cracks in our public policy, a focus on recovery. To do nothing, would be like "getting punched in the mouth but doing nothing to defend yourself." "Can you imagine if we treated deaths caused by coastal storms as we do those caused by terrorism?"

Merrell told me the cost of the Ike Dike would be $6–8 billion, "depending on how pretty you want to make it." But recent estimates put the cost at double that. A month after my trip, state senator Larry Taylor, R-Friendswood, filed Senate Bill 2265, which established the Gulf Coast Waste Disposal Authority as the responsible party for driving forward a proposed coastal spine. It asks Congress for $11.6 billion to fund the project.

It would provide protection not only to the island but the ship channel, and the East Harris County Manufacturers Association has endorsed it. The ship channel is fifty-two miles, and a storm surge can intensify as it funnels up it. Part of the design includes gates or barriers, similar to the one in New Orleans, that prevent surge from getting into the Galveston Bay. There, under the right conditions, a surge could intensify in the wide, shallow basin. A barrier would extend across Bolivar Roads, the strip of water enclosed by the northern and southern jetties between Galveston Island and the Bolivar Peninsula. The coastal spine would extend the existing seawall westward about eighteen miles to a point past San Luis pass and then eastward about thirty-five miles across Bolivar Peninsula to near High Island.

Merrell saw the protection more in terms of the need for industry and commerce, to protect people and property, than to prevent environmental damage in the channel, but either way, "it is in our national interest to protect that area." According to their cost-benefit analysis, 35 percent of the fuel the nation consumes and much of our fertilizer comes out of the area the Ike Dike would protect. The New Orleans levee system and barrier was $17.5 billion and an "eighth the size of what we're trying to protect, and a tenth of the economy." The population in the region is three times the size of that of New Orleans. According to George P. Bush, "We think it's sensible policy for the national government to be a financial partner. Bush did not say who the other partners would be, such as the industry that would benefit from storm protection.

Merrell, who sat in a leather chair, is small in stature but a big competitor. He mentioned fighting back when a storm hit, and twice in our conversation he stated, good naturedly, that "we won." Newspaper reporting on community meetings about the Ike Dike said Merrell gave little credence to other views. In our meeting, he cited the recent endorsement by Bush and the thirty or so coastal cities that supported his plan.

The competing idea that came from Blackburn's SSPEED center showed higher surges that account for climate change and greater storm intensity. They proposed a series of small structures farther up the bay at half the cost. With their hurricane predictions, Merrell thought they were trying to scare people, bringing up a Frankenstorm. Experts at Jackson State do his surge modeling, and he noted that they are a Homeland Security Center of Excellence. Merrell called them "greybeards," all retired from the Vicksburg

Corps of Engineers research center. Merrell himself is a physicist, used to modeling wave action and ocean processes before he became an advocate of the surge barrier.

Their modeling showed that a coastal barrier would have prevented almost 90 percent of the losses during the 2008 Hurricane Ike, with damages of $30 billion, the fourth costliest natural disaster behind Sandy ($75 billion), Katrina ($108 billion), and Harvey ($180 billion). Ike hit on September 13. That fall might have been the ideal time to ask Congress to fund comprehensive coastal protection. But on September 15, Lehman Brothers filed for bankruptcy and the fed stepped in to save AIG. Nature's fury took a back seat to financial fury.

I brought up the environmental impacts of the wall and gates. Merrell said part of the structure would be on the beach. "You won't be able to tell it from a sand dune." It would be seventeen feet high and made of different types of materials, some clay, but an outer covering of sand and seagrass to blend in with the surroundings. The Dutch walk on some of theirs. The footprint would not be a problem, he thought. Perhaps the gates that open and close to allow ship passage would provide an impact as they restrict the flow of water. But he said the ones in the Netherlands have not harmed their ecosystem, and there was even some modeling that some restriction could be good for the bay. As of now, it is starved from lack of fresh water ever since the Corps of Engineers moved the outflow from the Atchafalaya basin. Restricting some saltwater could restore the balance, be good for oysters. Then again, said Merrell, "it's not going to make everybody happy."

Sebastian Jockman, a civil engineer at Delft University in the Netherlands working with Merrell, told me the Eastern Scheldt barrier reduced flow in the estuary by as much as 70 percent. However, he thought the proposed barrier design for Bolivar roads would reduce flow by much less. Twenty percent is their current estimate, but this would need to be verified "with hydraulic and environmental experts and models."

When I brought up climate change, Merrell said that will be factored into the design. He talked of "adaptive management," that it is built for now, and if the sea level rises more than projected, design the structure so it could be built onto or reinforced. But no mention is made of climate change in the promotional video or in pleas to the president.

Merrell said that Bush would meet with President Trump soon, that there might be some infrastructure money for the Ike Dike. He hoped they would fast-track it, do a design-build as they did in New Orleans where procurement goes through a single point of contact and environmental review is streamlined. It took them five years to complete their system. Contender Merrell thought they could complete theirs in four. A month after my visit, Bush sent a letter to the president, signed by twenty-one coastal mayors, six county judges, and more than two dozen business people and educators, including Merrell, endorsing the project. The cost, which includes other projects in addition to the Ike Dike, had grown to $15 billion.

After my meeting with Bill, I drove west toward Follett's Island. Along the beachfront, they were piping sand and slurry from the Ship Channel into areas that had eroded. Once the new beach was created, they added more pipe and moved on to a new section, moving it around with dozers. The replenishment project had been going on for eighteen months at a cost of $20 million, a million cubic feet of sand between Twelfth and Sixty-First Streets. It was coming out as if a geyser. For tourism in Galveston, sand could be worth as much as crude.

A sign near Jamaica Beach noted that the area was a "campsite" for the Karankawa people, an indigenous tribe who roamed the bayous in dugouts, eating fish and shellfish when they could but moving inland when seafood was unsafe to eat in the summer months and later when tropical storms threatened. They were heavily tattooed and made an impression on Europeans. Men pierced nipples and their lower lip. The marker said they were "known for tall tribesman and ceremonial cannibalism." However, Cabeza de Vaca (yes, head of a cow), a Spanish conquistador who lived among the Karankawa in the 1530s made no mention of cannibalism. In fact, he acknowledged that he and his fellow Spanish conquistadors committed acts of cannibalism of their own to stay alive after being marooned off Galveston Bay. He wrote that the Karankawa people were shocked by the actions of the shipwrecked Spaniards. "Had they seen it sometime earlier, they surely would have killed every one of us." Shortly after contact with the Spaniards, half of the tribe died of cholera.

The marker said the population declined because of disease and warfare. However, when arriving in Galveston in 1823, Steven F. Austin, "Father of Texas," wrote in his diary that these Indians would have to be "exterminated."

In the Skull Creek and the Dressing Point massacres later that year, they were—or were chased off. The accusations of cannibalism might have been drummed up to convince settlers the Indians would be impossible to live among. According to David La Vere in his *Texas Indians* (2003), no reliable eyewitness verifies cannibalism nor does archaeology support it. If cannibalism was practiced, as among the Caddos and Atakapas, it was likely as a ritual, as a way of celebrating victory over enemies and denying them an afterlife, and not as a dietary preference or means of survival.

I drove over San Luis Pass onto Follett's Island. I stopped at the county park to watch fisherman bring in oysters from the bay. Galveston Bay is a productive estuary offering up catches of shrimp, crab, and oysters, but it is polluted, full of polychlorinated biphenyls (PCBs) and other chemicals, and consumption advisories are often posted. Nearby, Christmas Bay has better water quality and that is where the day's catch had come from.

I drove down Bluewater Highway looking for a spot to get out to the coast. Texas is among the states with the least amount of public land. I finally found a small park and bird sanctuary. I read a sign about the bird migration, how when the buntings, grosbeaks, flycatchers, and warblers came through, they resembled "smoke over water." I now had Deep Purple's "Smoke on the Water" stuck in my head. As I walked down the boardwalk and steps to the beach, lost in the three-chord progression, I was nearly struck by an SUV I did not expect to see roaring down the beach. Cannibalizing it as a road.

Despite beach traffic, this part of the island was far less populated. There were fewer houses, most high on stilts. I did not know if I was outside the length of the wall. What got put in and what was left out of coastal protection was still a matter for Texas and U.S. policy. I thought of the Dardar brothers on the Isle de Jean Charles, left out. Or the Karankawa, if they were still alive, only camping on the island, moving off during storm season.

John Anderson told me that Follett's was retreating at four meters per year. For now, there was enough room for at least two lanes of trucks. The erosion on the barrier islands was a function of at least two forces: 1) sea level rise and 2) lack of outflow. The Corps of Engineers moved the crow's foot from Atchafalaya basin, which meant these islands no longer received a refill of sand and sediment. A third factor was the jetties extending out from the ship channel, creating a safe harbor for boats but transporting the sand out to the Gulf of Mexico rather than down the coast. In these three factors, the law of unintended consequences seemed to be at work: solve one problem but create another. Dam the river but lose its creative power. Bring in the ships safely but carve up the beach. Pull oil from the ground but seal off the sky, heating up oceans and bays, supercharging storms. Would the Ike Dike have such impacts?

I drove back to Houston that night through traffic. The rodeo was in town so the streets were clogged with those heading to livestock shows and concerts. Houston sheltered some 150,000 people during Katrina. I once saw a demographer from the University of Georgia, Matt Hauer, give a presentation on where people might have to move given how sea level rise projections could inundate coasts. Hauer looked at IRS tax data to determine county to county migration. In other words, where people moved from or had relations. If sea level rose 1.8 meters by 2100, as some models predict, he expects Texas to pick up as many as 1.5 million people, mostly from Florida and Louisiana. Austin and Houston could net 250,000 sea level rise migrants each. Already, Houston has grown by 2.7 million people since 2010.

In the morning, I picked up the campers so we could return home. They had caught a few fish and chatted up some locals, but they were tired. Feral pigs were rooting up leaves outside their tent. According to Nathan Snow at the USDA, the range of these wild hogs is expanding about six miles a year. Due to milder winters, little piggy may go north. Yet another good idea gone bad.

I took a picture of the spring breakers before and after. Before, they are goofing around, sticking a head through the arm hole of the life jacket, making funny faces. After, they are smiling but more solemn. A couple of days in the woods always does it. Humbles you. Wilderness gothic.

Much is made about the lack of resilience of their generation, but they came out of their adventure fine. And if they feel entitled—to a viable planet, to protection from harm, to respect from all—aren't we all?

As they slept, I drove on the back roads up through Mississippi near Natchez. Behind the levees, fields lay flat and stretched to rows of trees. The sun was setting slowly behind us. There were small graveyards in the towns and white clapboard cabins up on cinderblock pylons, rusted farm equipment huddled under pines.

Wars have been fought here, not only to preserve the Union but to control the flow of nature. One is over. The other continues. I thought of another series of maps, those produced by Howard Fisk of the lower Mississippi River valley. Part of an otherwise technical report for the Corps of Engineers in the 1940s, the maps show traces of the former river in bright pastels, color-coded for time, a twisting knot of meanders, as a firehose when released. John McPhee wrote that the river "jumped here and there within an arc about two hundred miles wide, like a pianist playing with one hand." It has frequently and radically changed course.

To use another musical metaphor, if the Mississippi Delta were a guitar, the Atchafalaya and Mississippi Rivers would form the body, and we were approaching the frets. The sprawling course of the river depicted on Fisk's maps was laid over time through folds of the soil. The twists of those lines brought to mind the straight ditches and canals on maps by the Isle de Jean Charles, pipelines laid by blunt force. Rivers in their natural course follow the path of least resistance; pipelines follow profit.

From the radio came the blues, a show called *American Routes*. Songs about love and survival in the face of oppression, even violence, and heartbreak. It was resilience. Even if the river or hardship came, sang the song, it'll be alright. The sun will shine on our backdoor someday. But there's a kind of resilience like a river ecosystem, which needs the change. And there's another kind, through feats of engineering, which tries to avert it.

When our species grew a backbone, our ancestors gained the ability to breathe, walk on land, and emerge from water. Now, a coastal spine may be necessary to prevent some of us from retuning to it.

Postscript

As I was preparing my manuscript for deadline, Hurricane Harvey stalled over south-central Texas. As the flood waters rose, the rain continued to come down, as much as fifty inches, the most ever in U.S. history from a single storm. Houston received more rain in a few days than most states receive in a single year. Stories both harrowing and heroic were still unfolding as people posted videos of impromptu rescue efforts of the "Texas Navy," private citizens with raised pick-up trucks or bass boats or other watercraft patrolling, Dunkirk style, the swamped streets to bring relief.

The George R. Brown Convention Center, used to house Katrina refugees twelve years to the day, now housed the city's own residents.

The National Weather Service sent out this ominous warning on Twitter before the storm: "This event is unprecedented & all impacts are unknown

& beyond anything experienced." In other words, the storm was breaking both their scientific models, in that it could not be predicted, and previous weather records. Unprecedented, but was it to be unexpected?

The exact role climate change plays in hurricanes is hard to pinpoint, but this much is clear: our emissions are heating up the planet. The refineries and oil production that made much of the region's wealth, that made possible Houston's sprawling suburbs, played a role in this warming. And that warming helped lead to the bathtub-like conditions in the Gulf of Mexico, with warmer than average temperatures over the winter. And the warmer the gulf, the greater the amount of moisture available to fuel rainfall.

Kenneth Trenberth, a senior scientist at the National Center for Atmospheric Research, has said that "the human contribution can be up to 30 percent or so of the total rainfall coming out of the storm." Climate change "amplifies the damage" of any storm "considerably."

Houston has seen at least four one-hundred-year flooding events since the spring of 2015, according to the meteorologist Eric Holthaus. Some said Harvey rains were a once in a thousand years in terms of probability, but such events seem to happen both with more frequency and intensity to make such statistics meaningless.

Over the last twenty-five years, the city has lost almost 50 percent of its soil-rich wetlands to development and concrete. Wetlands naturally mitigate flood, absorbing rainfall. According to a *ProPublica/Texas Tribune* investigation, more than seven thousand residential buildings have been built in FEMA-designated flood plains since 2010 in Harris County, which includes Houston. Rampant development and its attendant roads, roofs, and parking lots, without adequate preservation of wetlands, likely exacerbated the flooding, as did climate change. Without zoning, the metropolitan area was poorly adapted for climate change, an example of how not to build a climate-resilient city. As some have said, parts of Houston drowned in freedom from regulation.

Writing in the *Washington Post*, managing editor for the *Houston Chronicle*, Vernon Loeb, writes that Houston should use the storm to reinvent itself. The city should "use Harvey to jump-start its transition from the country's epicenter for oil and gas to a world capital of alternative energies."

In the days and weeks after the storm, while people rebuild, the response and plans will be evaluated, arm-chair quarterbacked, particularly the decision not to evacuate. The mayor of Houston, Sylvester Turner, said the decision was based on what happened during Hurricane Rita when a hundred-people died stranded on the roadways. But bad decisions then, and the right lessons learned, need not mean bad decisions in the future. Why was there no plan for contraflow (reversing traffic lanes) to ease congestion? For evacuating the elderly and sick? Why was there no place for people to go?

Whether or not Hurricane Harvey is a game changer remains to be seen, but just weeks before the storm, President Trump rolled back commonsense requirements for flood standards for federal projects. Much will be made of the response of first responders, and how the city will come back stronger and indeed, the cooperation of residents in this diverse city was encouraging. But whether or not the conversation shifts to preparedness rather than response, to dealing with climate change rather than denying it, as we have for nearly forty years, remains to be seen.

Perhaps, at long last, we will at least turn our attention to a different kind of wall, one holding back storm surge rather than people. Such a wall might not have helped in this particular storm, but that storm is coming, in Houston or somewhere else along the coast, where there will be billions more in damage, and we will eventually get tired of bailing out.

Our planet lived in a kind of temperature equilibrium for thousands of years. But more recently, warmer temperatures have disturbed that balance, ratcheting up extremes of rain and drought, fire and flood, launching out of any normal orbit. We are out of the Holocene, the period after the Ice Age, when temperatures became hospitable to most life, and into what climatologists call the Anthropocene, when conditions on the planet have been dramatically altered by humans. The craziness of these storms, and the damage they bring, could be the new normal, and something we will have to get a grip on. If not, we might as well waive a white flag and retreat for higher ground. But that seems unlikely to fly in Texas.

8

Take in the Waters

On the Birthplace of Rivers, West Virginia

At a convenience store, I picked up a copy of the *Market Bulletin*, put out by the West Virginia Department of Agriculture. It contained articles about the state fair and profiles of farmers and lots of recipes for pumpkin: pie, cookies, cornbread, and pudding. In the back are ads for boats, hay balers, chickens, and other things important to rural life. There was also this headline. "Severe Wet and Then Dry Weather Dampen Harvests." A dry April gave way to a dry May it says. "That was followed by a June and early July that felt more like monsoon season." The crops Andy Crihfield put in the ground did not fare well. "My father has been farming since he was 13. He's 85. This has been one of the wettest summers he says since he's been farming." But Crihfield wasn't crying into his corn. He said the late crops they planted in July and August look good. Green beans did especially well this summer. "We're just going to thank God for what he gives us. And hope that next year is better than this one."

I was headed to a conference in West Virginia's Allegheny Highlands at Blackwater Falls State Park in Davis. Scientists, public-policy experts, lawyers, and environmental activists would be gathering to focus on climate change in the highlands and what citizens and leaders could do about it.

West Virginia's Corridor H is a ribbon of road that slices through ridges and crosses the rivers and valleys, affording far-reaching views. Some folks on both the right and left call the road "Robert Byrd's Road to Nowhere," after the former senator and coal miner and his affinity for what some see as pork spending.

At the top of another ridge, giant wind turbines turned slowly in the wind, majestically. And yet, they have changed the ridge forever. From a distance, the white blades fade into the clouds and are unseen, yet they are also always there. At my exit was a different monstrosity, the Mount Storm Power Station owned by Dominion Resources, their biggest coal-burning facility. Smoke stacks scraped the sky, releasing plumes into the wind.

Inside the conference venue, the conversation is all about moving away from a coal-based economy, but that won't be easy.

Tom Rodd, conference organizer and director of the Allegheny Highlands Climate Change Impacts Initiative, talked about how climate change and social justice are intertwined, the way low-income communities are affected by pollution. But it can be very difficult, even for those affected, to extract themselves from an economy of extraction. "Nobody likes to hear the work they do is destroying the planet," he told the audience.

Part of the solution, he said, was finding the transition away from this economy, replacing it with a new one. What is at stake in doing so? He and other presenters talked of the loss of fish and forest, increased flooding and heat. Most recently, he said, he has been speaking to teachers and students rather than politicians, but "everyone is adjusting." Rodd said his generation created the problem, but the next will have to fix it. "We didn't know what we were getting into, we left you guys a mess, but here's what we're going to have to do." When I later called to talk to Rodd, he touted the work of the state's science teachers who invited Lonnie Thompson, one of the world's foremost authorities on paleoclimatology and glaciology, and a West Virginia native, to speak at their fall 2017 conference.

Education was Rodd's way to reach people, but education can have its special interests too. In 2014, Wade Linger, a member of the West Virginia state school board, suggested changes to the state's new science standards regarding global warming, modifications that brought significant local and national criticism. He wanted to add the words "and fall" after "rise" to the standard requiring sixth graders to "ask questions to clarify evidence of the factors that have caused the rise in global temperatures over the past century," casting doubt on the overwhelming scientific consensus that human fossil fuel emissions are a driving force behind global warming.

Linger, a technology entrepreneur, told the *Charleston Gazette-Mail* that he wanted to add "and fall" because of research he had done showing temperatures rising and falling. He called global warming a "binge," and wanted a discussion of Milankovitch cycles, the long-term variations in climate that show temperature rising and falling with carbon dioxide levels.

At the conference, Amy Hessel, a professor of geography at West Virginia University, said that while temperatures have risen and fallen, the overall trend of late is clear. When she talked about Milankovitch cycles with students, she used them to show the human effects on climate.

Hessel took us on a "time machine," looking at tree ring cores in a species of Siberian pine that can live two thousand years. These "archives of climate data" told of drought during the time of Genghis Khan and his raids on eastern Europe. Drought likely led to a search for available resources and conquest. Climate change presaged other ecological and social changes.

To see social changes and economic transition on a more recent scale, including a proposed wind farm in coal country, I drove about an hour from my house. East River Mountain Overlook is named for the river that originates in West Virginia but flows east. On a warm summer day, the sky was clear and the wind scarce, though it often howls on that high ridge, with a view of the towns of Bluefield, Virginia, and West Virginia below. No one else enjoyed the view with me, though just before I pulled into the parking lot, a young bear paused in the middle of the street and watched me with curiosity.

At the overlook, a new development across the street, with lots mostly sold, some men were digging out a drainage ditch for the first home built, a log cabin perched on the ridge, porches on both sides of the house. Watch the sun rise and set. With a giant claw on their dozer, they were attacking a particularly stubborn piece of limestone that would not budge. The rubble gave up easily, but now, they had a grip on the spine of the mountain itself.

The owner came out, an older woman carrying her morning mug of tea. She was a retired nurse. She had a house at the beach and one here in the mountains. She preferred the mountains, "because I'm older, and I don't like people." Her kids thought her crazy, "there's nothing to do here," but she liked it that way. She had emigrated from Vietnam. To afford both places, "I worked hard. This is America, right?"

I asked the day's question, what she thought of the wind farm that would go up on the ridge south of her. She had not heard of it but wanted to know how she could get one on her place because the wind did blow. I told her the wind farm was still controversial, that it could be blocked, but she thought if she put one on her property, what could they do? I did not want to say what they probably could do, restrict though zoning or other laws. Then again, if they let houses on the ridge, why not a wind mill?

On the streets of Bluefield, West Virginia, restaurants and buildings were for sale. It was hard to find much activity.

I saw a man heading into an auto parts store and asked him what he thought of the wind farm. "I have no problem with it. Especially if it brings down my electric bill." But he started to reconsider. Would it be bad for the mines? He was a security guard for a mine over in Princeton, West Virginia, working the night shift, keeping people out of the mine. He would not want anything or anyone to come in, including a windy competitor.

Two men pumping gas into an old blue truck, a father and son, had heard about the wind farm, had no problem with it, and wondered if the logging that had been done at the base of the mountain was related to it. I drove over to see. Indeed, a large swath across the base had been shaved close.

The town hall of Bluefield, Virginia, was near the mountain, so I walked in to inquire. Dee Cox, the receptionist, told me it was logged by a private timber company and admitted it looked ugly. She pulled in the zoning administrator who told us it was for a development, housing and condos, Leatherwood. I asked about the wind farm, but he had no news. I asked if he knew why a community could stop a wind farm but not a pipeline. He did not have an answer but thought it had something to do with eminent domain, the ability of governments to take land in the name of "public interest." One of the puzzles of our national energy policy is why this could be done for pipelines but not for other forms of energy.

Dee Cox said there would be a story that evening about a pipeline, and that I should pop over to WVVA, the local TV station. Sure enough, they buzzed me in and took me into the room with surround TVs, a live newscast going on from the Walmart and what looked to be a Yul Brynner western on a sister channel where a woman falls into a cabbage patch. The news story was on a four-hundred-acre historical orchard farm, Doe Creek, in the route of the proposed Mountain Valley pipeline. The owner, Georgia Haverty, planted hundreds of dwarf apple trees, replacing the old orchard so pickers would not have to climb ladders. There is a smokehouse on the property from the 1880s, rustic and historic. They have started to host weddings in the renovated pack house. The basement is a bar with old cider casks. The Federal Energy Regulatory Commission had recently released its final environmental statement, but they found little to halt the pipeline's progress. The company behind the project has designated places like Doe Creek Farm and the nearby town of Newport, Virginia, with a covered bridge a "high consequence area" in case of an explosion. That means they are eligible for a higher yet no less comforting grade of pipe. The Hazardous Materials Safety Administration prefers the phrase "potential impact radius." Others call these areas, often poor, a "sacrifice zone."

At the New Graham Pharmacy, The Last Fountain, a classic lunch counter with a BLT and macaroni salad for under five, I asked the doctor to my right, still in scrubs, if he knew about the wind farm. He did not. Kenny Thompson, to my left, said "everyone was against it," but that was not my experience so far. He said it would "tear up the mountain." I noted the logging had done that pretty well too, and Kenny, with long hair, taking a pause from his burger, his gun holstered to his belt, noted that they too, "tore up the mountain." "Bunch of lawyers and doctors gonna live up there," he said. The waitresses behind the counter had overheard. What are they doing to the mountain? The manager or owner had been barking orders at the staff, "register," get on that BLT, and they had been responding with "yes ma'am," a tight ship. But she took a break from the grill, "where's the animals supposed to live?" I brought up the morning's bear.

The swath cut through the most biodiverse temperate hardwood ecosystem in the world, a relic of the ancient woodland that once covered much of North America. Majestic sugar maples, hickory, oaks, and tulip trees towered above the cut, hugging the steep slopes of the Appalachian Mountains.

The state line is something of a mystery here. The town of Bluefield, Virginia, voted to adopt the name of its neighbor, Bluefield, West Virginia, in the 1920s, though the Virginia city is west of the West Virginia one. A ceremony was held as was a symbolic wedding with residents from both cities, attended by the then governors of both states. The governor of Virginia was the best man. West Virginia's governor gave away the bride. The name is derived from a species of chicory that has a periwinkle blue flower, the root of which is used in coffee at the Café du Monde in New Orleans. As any lawn mower knows, the stem is woody, hardy, and hard to take down, like the people in the region.

When I reached Tazewell County, Virginia, supervisor Charles Stacey by phone, he told me that Dominion still has plans to build the wind facility.

They own about 2,800 acres on the mountain. And the county was still opposed to it. The county passed a "tall structures" ordinance, limiting buildings above 40 feet, but Dominion has not requested a variance. The main reason for the opposition was aesthetic, he said, saying that the impact of building on the "critical ridgeline" would be like putting a four-lane highway up there. But he also acknowledged a second reason, having to do with the deep ties in the area to coal, a competing industry for energy. Because coal was an "industry under siege," it was a bitter pill to see a wind farm go on the mountain that would at most supply ten new permanent jobs. Stacey said he could not see the mountain "desecrated" in the name of something that would further harm the industry so many in the area have relied on. "Coal has been our livelihood."

Stacey, an attorney and Democrat, said he felt like a *Washington Post* story by Jenna Portnoy, "After Coal: Appalachia to Wind," represented the issue as "hillbillies don't know what's good for them." The article quoted a blogger, Lowell Feld of Blue Virginia, who pointed out the hypocrisy of hating wind turbines but blowing up mountains for access to coal seams. Many of the commenters jumped on this. Stacy told me that he and other officials were not against renewables, though he did not think the wind farm would go up if Dominion was not looking at subsidies or other carbon off-set credits, referring to the wind farm as a "feel good" solution. The proposed farm would include about sixty-four wind turbines. He would be in favor of an even bigger project, two thousand of the things at a different site, or a giant solar farm on some reclaimed mine property, but Dominion has "refused to come down to look at the sites."

In regards to the mountain aesthetics, Stacey said the wind farm "might not hurt but doesn't help" efforts to promote tourism and a network of all-terrain vehicle (ATV) trails. When I asked about doing something about climate change, he acknowledged the reality of it, though many of his constituents doubt the science and that the situation on the coast was more dire. But that did not mean Tazewell had to sacrifice their mountain. He recognized, too, that "change is hard" for people and that they were going through their share of it in the county.

Leaving Bluefield, I drove northwest to Welch, West Virginia, the county seat of the southernmost county, McDowell. The drive out of town on U.S.

58, the Coal Heritage Highway, is dotted with pawn shops, car washes, and makeshift flea markets by the roadside. I passed through whole towns that seem to have disappeared, Switchback and Powatan, though I learned that the floods of 2001 and 2003, when the Elkhorn and Tug Forks overflowed their banks, did them in as much as the loss of the coal economy. Once proud schools sat empty. In North Fork, signs told of state championships in basketball through the late 1970s and early 1980s, but it does not look as if they could field a team anymore. On the hillside, houses are swallowed in vines, windows smashed.

In Elkhorn, the Clark National Bank is boarded up, but there is a plaque honoring Elizabeth Simpson Drewery, the first black woman elected to the West Virginia Legislature in 1950. Her family was part of the great migration into the coal camps, and she taught in the schools in McDowell County. In the legislature, she exposed a bribe on the part of the coal companies and advocated for labor, education, and civil rights. Miners often worked side by side with immigrants and people of color.

I stopped at Five Loaves and Two Fishes Food Bank in Kimball, just before Welch. Linda McKinney, the director, told me of the donations they get from far and wide. They used to receive some from the nearby Walmart, but it closed the year before. She gave more grim statistics. Seventy percent of children in the county live in homes with parents who do not work. Nearly 50 percent live in homes without a biological parent, raised by grandparents or other family or foster homes. McDowell has the lowest life expectancy in the nation for men, second for woman. What do they die of? According to some, despair. Because there is little worthwhile work or future to look forward to, many turn to substances to dull pain and pass the time, which leads to a downward spiral. "I'm the only laugh, hug, or smile they may get on a given day." People come on the third Saturday of the month to get donations of canned food or toiletries. Often, they sleep in their cars in the parking lot the night before.

Outside, her son Joel worked on raised beds. He went to college, dropped out of it, and joined the navy. It straightened him out some, but afterward, he partied a lot until he got a DUI. He went home to serve a ninety-day suspension and started thinking more seriously about what he wanted to do. Working in the mines was not it. His father did that. At thirty-three, he enrolled at Penn State, earning a degree in agriculture business online, while keeping the greenhouse, which he called a tunnel, going.

He began a GoFundMe campaign to do a community garden, create a local food economy, and it has worked. After three years, he has finally turned a profit. He grows everything hydroponically, because of mine run-off in the soil. "It's just a hot mess," he told me. Bok choy and kale grows out of containers, towers he calls them, freestanding buckets stacked on top of one another with holes for the plants to grow out of.

Linda had told me their site is a homeland security disaster meeting place. There is no Salvation Army or Red Cross nearby. They are it. And during the election, Bernie Sanders came through and held a rally, only he was not supposed to be "political," as they are a nonprofit. They advocate for food, for people, but not politics per se. He came again after the election and MSNBC covered it. When they spun it as "Bernie in Trump Country," Joel said he lost it. And he grew further pissed because they were supposed to cover poverty, as everyone does, opioids, that story well-known too, but also something like the positive changes going on. Instead, they ran B-roll for that part, another story about drugs.

"Just because people did not go to college does not mean they are not intelligent. A lot of these people have skills, talents." The schools in impoverished communities do not prepare people for college, he said, and the colleges do not prepare some of these people for "real life," or at least labor force training.

I asked about a proposal for a new Homestead Act, floated by Jim Branscome, of buying back acres from bankrupted coal companies and releasing some federal land holdings to private individuals. Joel said it was true that people wanted land, but he thought it a case of "too many chiefs and not enough Indians." Plus, as much as 60 percent of the land in McDowell County is owned by private companies, mostly coal, waiting to harness timber or minerals. Ron Eller, a professor of history from the University of Kentucky, has written about the legacy of that ownership system, that some form of "long term leasing of land in a rehabilitated Appalachian Commons could provide opportunity and hope for another generation of local entrepreneurs and give new meaning to the term 'homesteader.'"

Many had told me that if I wanted to find some folks doing positive things for economic adaptation I should talk to the Coalfield Development Corporation. So I spoke with Ben Gilmer, president of Refresh Appalachia, one of the social enterprise spin-offs of the organization.

When it started, the Coalfield Development Corporation provided education, on-the-job training, and personal development to help rebuild the struggling West Virginia economies from the ground up. They worked with the unemployed and underemployed, building or renovating low-income housing. It was run like a business with a non-profit-like mission, so a social enterprise. They wanted to help former coal workers not just swing a hammer or learn jobs skills but help them in more holistic ways, allow people to reach their full potential.

They had a thirty-three, six, and three model carved out of a forty-hour week. Thirty-three hours were spent working, where they were also trained, six hours at a community and technical college, and three hours in personal development, which could include lessons on personal finance, parenting, and health. Ben related a story of one person who made a pitch for building cabins on his land along the Hatfield-McCoy ATV trail. After the pitch, which everyone loved, they asked for a business plan. "I just gave you my plan." He did not understand that the bank wanted a more formal proposal.

The Coal Development Corporation now applies that same model for job training to other sectors. Rewire Appalachia is focused on energy such as solar installation. Rediscover Appalachia focuses on tourism and local business. Refresh Appalachia, which Ben is in charge of, focuses on food production and systems. They get a mixture of federal funding and private grants but try to be as self-sustaining as possible. Initially, they worked with eighteen- to twenty-four-year-olds, but increasingly, those in their mid-forties came in, used to making $80,000 to $100,000 in the mines.

At Refresh, they are trying to create not only local foods but improve access to it. They focus on the production but also food hubs, where they aggregate, package, and distribute the food. Lately, they have tried to create microgroceries in convenience or dollar stores, food deserts of sorts. "People want to do their part, buy this stuff if it was grown locally." He said that in one program in a high school, some of the kids grew purple carrots. Some of them had never had carrots, much less purple ones, but soon they were "tearing up the salad bar because they grew it."

Ben has a background in geography and environmental studies. On a given day, he wore many hats and could have to both fix the hydraulics on a tractor and meet with a private foundation from Pittsburgh. He once worked with the Nature Conservancy in climate change adaptation and could foresee a future where the food production shifts east. However, "we have to make something out of what we have, which is not premium ag land." Much of the land is hilly, or forested, or reclaimed from a mine. In one of their farms in Mingo County, they have an orchard on three hundred acres of a former surface mine. There, they grow native paw paws, an Appalachian delicacy, berries, and lavender. To help rebuild the soil, they have goats chew up some of the invasives, chickens to scratch and add nitrogen to the soil, and then pigs to add more nutrients. Put some wood chips on top, some carbon, and the scraped-off mine starts coming back to life.

He also said they are trying aquaponics, which is the combination of aquaculture (raising fish) and hydroponics (the soilless growing of plants) to grow both fish and plants together. They are using geothermal from an abandoned mine for energy, keeping the system from freezing in winter, cool in the summer. Ben mentioned some "premium water" locked up in mine shafts, depending on the site. And how a county just across the state line in Virginia used pump storage from a mine shaft for energy to cool a data center. The water in those shafts could be among the many "assets that haven't been unlocked that will be critical with climate adaptation."

I asked Ben about recent success stories, and he said that that many of the people, though they still felt a strong cultural identity with coal, were "just excited that something was going on. People around here are innovative. They fix stuff. They see a solar panel going up and they say hey, 'John has a job, and look at that cool new thing on the roof.'"

The biggest obstacle Coalfield Development Corporation faces is a psychosocial one, of overcoming hopelessness, "generations who have not been empowered to imagine alternative futures." They assume that they will follow the accepted, traditional path, which usually leads to the mines. "We provide something other than the usual path," said Ben. "Having an alternative pathway is the key to adaptation."

I had been to Welch before, once with my daughter, Elliot. I have often taught *The Glass Castle*, which college students respond to, and Elliot enjoyed the book as well. Jeannette Walls's memoir begins out west in California and Arizona but takes a dark turn when the family arrives in Welch, father Rex Walls's hometown. Though his parents still live there, both Rex's drinking and the Walls family poverty worsens. The house they move into has no running water and sporadic electricity, a hole in the roof. There are hints that the father suffered from sexual abuse. But Walls, an electrician and engineer, dreams of building a glass castle, a family palace, passive solar. In Welch, in the backyard of the family's house, in a pit where the castle might one day be built, the family begins to store their trash, a metaphor for their ruined dreams. We sense the move will not go well as soon as they cross the Tug River, which Rex explains is so full of fecal matter that one cannot swim or fish in it. To Jeanette, the entire town seems to be covered in coal dust.

During our trip, Elliot and I asked at the local library for someone to talk to who knew the family. They sent us to a woman, Priss Freeman, at city hall, whose mother took care of Maureen, the youngest and somewhat neglected daughter. She had not read the book but had not heard good things about it. Several people we spoke to said the same. They thought the book represented the town in a negative way, when it is more a profile of a dysfunctional family than a town, though even in the late 1970s when Jeanette left for New York, Welch was on the decline.

To learn more about the town's history, we spoke with Danny Barie, a lawyer. He said the population peaked at about one hundred thousand in the fifties but was twenty thousand now. Ten thousand was sustainable, he said. The decline began when metallurgical coal began to decline. Met coal, abundant in the area, was used for steel not steam, and when the steel industry declined in the United States, so did coal in the region. Billions came out of the area, but not a lot was invested back in. The profit came out, the pollution stayed behind. Because of the destruction caused by coal, no other industry has moved in. There used to be two movie theatres. The last film listed on the marquee of one of them? *Born for Trouble.*

Barie told us that two most notorious events in town were the shooting of Sid Hatfield (a relative of the original Hatfield and McCoy feud) on the courthouse steps in 1921 and the publication of *The Glass Castle*. "The town has come to accept the former but not the latter." Sid was police chief of Matewan during the Battle of Matewan and was gunned down by Baldwin-Felts detectives (hired by coal companies) while awaiting trial for another incident. He was in favor of workers' attempts to unionize.

We drove on to find Hobart Street and the actual site of the house. The road is carved into the side of the hill, the houses against the bank. At the site, we saw little but the remains of steps. My daughter shared her journal for the day:

> Queen Anne's lace and black-eyed Susans line the Coal Heritage Road that was someone else's heritage before it belonged to coal. We're searching for what is left; broken people living in a shell of a town with more empty storefronts than full, pristine "friends of coal" stickers contrasting with the dusty landscape, as if they had no choice but to get in bed with the industry that left them battered and shaken. We confront the gaunt face of poverty when we cross up Little Hobart Street, the exact place Jeannette Walls' house stood years and years ago. The man who speaks to us about Welch continuously reminds us that he "speaks to us in very hushed tones." As if children are present who can't learn the truth.

At one time, coal offered promise. Bluefield was a bustling metropolis called "Little New York." In the essay "Wealth," collected in *The Conduct of Life* (1860), Ralph Waldo Emerson wrote about the promise of coal:

> Every basket is power and civilization. For coal is a portable climate. It carries the heat of the tropics to Labrador and the polar circle; and it is the means of transporting itself whithersoever it is wanted. Watt and Stephenson whispered in the ear of mankind their secret, that *a half-ounce of coal will draw two tons a mile*, and coal carries coal, by rail and by boat, to make Canada as warm as Calcutta, and with its comfort brings industrial power.

Coal was "portable climate" for Emerson, the fuel for civilization. But by the end of the nineteenth century, the Swedish scientist Svante Arrhenius had made the connection that the carbon dioxide caused by that coal could affect the climate. Arrhenius saw that this human emission of carbon would eventually lead to warming. However, because of the relatively low rate of carbon production in 1896, Arrhenius thought the warming would take thousands of years, and he expected it would be beneficial to humanity.

In Appalachia, some five hundred mountains have been leveled. Coal slurry ponds leak, poisoning streams, leaching arsenic, lead, cadmium. Tom Rodd, of the Allegheny Highlands Climate Change Impacts Initiative, said they are in a colonial situation where the minerals are owned by companies out of state. The people are coping with this "wrenching transition" where natural gas helps out in terms of jobs and revenues, but it brings the mines down. He referenced a public policy book called *Tragic Choices* (1978) by

Guido Calabresi and Philip Bobbitt. On the one hand, communities want jobs, revenues, schools, but on the other they are losing the planet. What do people do when going through the process of tragic choices? Like the first stage in the grieving process, denial is one very real option. But once we accept the state of things, we can begin to make suitable choices for dealing with the inevitable, to minimize suffering.

On the way home from Welch, I took a detour off the Coal Heritage Highway into historic Bramwell, a blooming oasis off the main road. It sits in an oxbow bend in the Bluestone River, tucked into a hollow with historic homes, mansions really, about twenty of them along the river and historic brick streets. They are Victorian charmers, Queen Anne style, gracious and elegant, too nice for the neglect affecting other coal towns in the region where the homes are built fast and cheap. These houses would run half a million dollars or more in most towns but probably sell for one-quarter of that or less in a former coal town that used to have 4,000 people but is now at about 350. I would learn that nearby Pocahontas, Virginia, housed many of the workers while Bramwell was home to the mine owners, some of the richest and most successful people in the 1890s coal boom. The town was said to once have more millionaires per capita than any other town in America. The school mascot is named for them, the Millionaires.

High-quality, hot-burning smokeless coal seams ten feet high began near Bramwell and extended for fifty or so miles. Speculators, engineers, entrepreneurs, and miners flocked to the village, near the very first coal mine in West Virginia in 1884. Many came from abroad, straight from Ellis Island. Black people came from former slave states. When all arrived, they lived in company-owned housing and bought their supplies at the company-owned store with scrip.

I was able to join the mayor of Bramwell, Louise "Lou" Dawson Stoker and her daughter, Dana Stoker Cochran, on the porch of Lou's home one summer evening. Lou, as charming and gracious as the town, was just elected to her sixth two-year term. She began to tick off the history of nearby buildings, including her own. Her house was once the Bank of Bramwell, though you would never know it by the clapboard siding. Inside are several old vaults. The current bank sits catty-corner from her house, an imposing structure that was once a financial focal point of southern West Virginia.

The National Register of Historic Places describes the Italian marble on the floor of the "two-story stone gabled structure with round arched doorway and transoms on front first floor windows." Across from it is The Corner Shop, a soda fountain now serving milkshakes, burgers, and fries to patrons of the region's new industry, all-terrain vehicle (ATV) riders.

As mayor, Lou made Bramwell a gateway for the Hatfield-McCoy Trail System created by the West Virginia Legislature to generate tourism and economic development. It covers hundreds of miles of off-road trails in six counties. Users purchase a pass, like a fishing license. The system is named for the feuding families near the West Virginia and Kentucky border after the Civil War. Many of the communities are "ATV-friendly," allowing visitors to ride into town.

I had seen the machines and riders, some so caked in mud they are indistinguishable from one another, stop at various pullovers in and around town. One of the first questions I wanted to ask Lou and Dana was if there were complaints about the noise or traffic. Lou worked very hard to get the trailhead in Bramwell because she thought it was important economically but there is a bit of a love/hate relationship there. However, the riders keep the buildings occupied and businesses busy, so they put up with the noise.

Guests can stay at her house, called The Bluestone Inn, though it has no sign. As we sat on the porch, we were joined by Melissa Sabo, manager at the larger Bramwell ATV Resort. They have rooms in one of the old post offices (when there were many in town), cabins, and a house that was an old coal company house. Melissa said they keep the places somewhat authentic, leaving the "seventies puke green carpet as is." They provide riders with grills and wood for a bonfire. Melissa has hosted guests from as far away as Alaska and Puerto Rico.

Lou has been working for over thirty years on economic development and historic preservation in Bramwell. She said that part of the reason the town was so well preserved is that the owners actually lived there. Today, much of the land ownership in nearby counties is by entities who are out of state, absentee ownership. Ron Eller, a professor of history at the University of Kentucky, writes that because of absentee ownership, "the dollars that could have built better schools and better roads and better health services in the early part of the century flowed out of the region, and we got what we call growth without development." Appalachia got boom growth, "but we didn't get the development of those aspects that will sustain a community over time and provide a quality of life."

Sitting in a wicker chair, Lou is a fount of knowledge on coal and Appalachian history. She knows the name of the cartographer who worked for Stonewall Jackson, Jedidiah Hodgkiss, who came through the area looking for mines and minerals. She told me who surveyed the Pocahontas coalfields, Captain I. A. Welch, namesake of the town in McDowell County, former resident of Bramwell.

Lou's own history with coal embodies some of the mineral's complicated mix of bestowing both sustenance and tragedy on people. Her father died when she was nine. Her husband operated a strip or surface mine, though she and daughter Dana were careful to point out how tightly regulated these were compared to mountaintop removal, where whole mountains are blasted away to access the coal deep within.

With a group that calls themselves the Millionaire Garden Club, though none of them are millionaires, Lou and volunteers listed the historic houses on the National Register of Historic Places. They planted dogwoods. They had the town registered as a bird sanctuary. "Why would people come here?" they kept asking. They held open houses, Christmas tours. With that money, they bought a police car. A sense of civic pride emerged.

In the 1990s, Lou was part of a delegation that lobbied for a National Park designation for Bramwell. She went to Washington, D.C., with Congressman Nick Joe Rahall. They thought Bramwell would tell the story of coal, the railroad, the mining camps, economic classes, union struggles, immigrants, integration—"we had it all." Opting in for property owners was voluntary, but the community supported it. Rahall introduced H.R. 793, the Bramwell National Historical Park Act of 1993, which "establishes the Bramwell National Historical Park, West Virginia, to preserve, restore, and interpret the unique historical, cultural, and architectural values of the town." The park service superintendent at the time told them it was premature.

Still, they got a renovated train station out of the deal. And the answer to the garden club's question, why would people come here, has come in the form of dirt bikes and ATVs, ridden high up in the mountains on former coal and logging roads and parked across the street from the old train station that once deposited newly minted miners. As an economy, ATV visitors will never replace coal, but it "helps keep the things steady in the meantime," Lou told me. "There are trade-offs," daughter Dana said, "the bad with the good, but we like where we live, and it's working out." Meanwhile, Lou is already on to her next project, a "blueway," like a greenway trail but one for kayaks and canoes along the river.

I kept traveling to West Virginia to find more stories of adaptation. To get there, I followed the course of the river. The road parallels the New River as it slices through the Appalachian Mountains. At a section pushed up millions of years ago, the river narrows. There, I turned right at Rich Creek and headed north through Peterstown, West Virginia, the town that nearly struck it rich.

In 1928 a boy was pitching horseshoes in his backyard. As luck would have it, he found a diamond.

William "Punch" Jones's horseshoe kicked up some dirt near the stake and revealed an alluvial diamond weighing 34.46 carats, the second-largest known in North America. The family was already famous for setting a record for consecutive male births—Punch was the eldest of seventeen. He kept the diamond as a curio in a shoebox during the Depression until he showed it to a Virginia Tech professor in 1944. It then went to the Smithsonian for display and was eventually sold at Sotheby's for $75,000 in 1984.

Geologists aren't sure where it came from, but speculation is that it washed downriver. Though a good mile from the current channel of the New River, Punch played in an ancient riverbed.

I learned some of this from a road sign as I traveled north on Highway 219, the old Seneca Trail, long a travel and trading corridor used by the Catawba, various Algonquian tribes, the Cherokee to the south, and the Iroquois Confederacy to the north. The road crosses many rivers: the North Fork of the Blackwater, the Shavers Fork of the Cheat, Tygart, Elk, Greenbrier, and Indian. I traveled this road to learn more about a special designation that would create a national monument to honor these rivers and more, to make a section of the Cranberry Wilderness and some of the Monongahela National Forest into the Birthplace of Rivers National Monument. It would preserve one of the largest stands of contiguous wild forest in the East and protect the headwaters of six rivers: the Cranberry, Gauley, Greenbrier, Elk, Williams, and Cherry.

I love to travel this road for the view, but it twists with the rivers, rises and falls with the topography. The geology that I've heard about this area says that when an ancient continent slammed into our own it smooshed up the mountain and some of that material fell back over the vertical, upside down, creating a series of folds and lumps. The road follows the valley and ridge topography, the elongated dorsal fin of Peters Mountain and the Appalachian Trail to the east, goose bumps to the west.

The road seesaws until I pass by Indian Creek, a covered bridge nearby and the remains of an old resort, Sweet Springs, one of the many places where people used to "take the waters" in the nineteenth century, a remedy for a range of ailments. The first five presidents all stayed there as did the Marquis de Lafayette. I was on my way to Lewisburg, near the Greenbrier in White Sulphur Springs, another of these spas. They were popular before air conditioning when people wanted to escape the heat and humidity by heading for the cool mountains, soothe their joints in the warm, curative mineral baths. The waters were said to have a quieting effect on the circulatory and nervous systems and offered a sound night's sleep. But their novelty eventually wore off, and now only a few remain, such as the Greenbrier or Homestead, which were favored by the railroad.

The local springs are connected with disturbances in the geologic structure of the area. Along the fractures and faults, groundwater heated by the Earth's core rises to the surface.

On the drive, I enjoyed the road signs: no pipeline, no fracking pipeline, no frack in karst, Permanent threat 4ever, Savethewatertable.org, PreserveMonroe.org. And yet, once solidly democratic, West Virginia went for Trump more than any other state, narrowly beating Wyoming, Oklahoma, and North Dakota—other energy economy producers.

There were Trump signs but also ads for Jim Justice with the words "jobs, jobs, jobs," featuring men in hardhats. "Tired of being 50th?" asks one sign, a reference to the state being last in several categories: finances, education, and health. Though two-thirds of voters pulled a lever for Trump, they elected Democrat Justice for governor (who switched to the Republican party in August 2017). In 2009, Justice, a coal and agriculture businessman, purchased the Greenbrier for $20 million, rescuing it from bankruptcy.

When candidate Trump held a rally in my town in southwestern Virginia, I went to see if he would say anything specific to the region. He mostly gave the standard stump speech until the end when someone asked him about coal. "Coal," he said, when prompted. "We're going to bring the coal industry back 100 percent." "Clean coal," he added, "I hear it works."

Most economists agree that the decline in coal jobs is due to mechanization in the mining industry and a market shift by electric utilities away from coal and toward cheaper natural gas. Telling the region that coal jobs are coming back has been likened to telling New Englanders that the whaling industry will return.

I wanted to visit Lewisburg because the film event being held there was trying to help envision a future beyond coal, to create a new economy. The "Live Monumental" film tour was put on by Keen, a shoe manufacturer, and several young people traveled the country showing films that highlighted a love for wild places and their support for national monument status. Among the Birthplace of Rivers were sites proposed in Owyee Canyonlands in Oregon; Boulder-White Clouds, Idaho; Gold Butte, Nevada; and Mojave Trails, California. The staff brought a young, outdoorsy vibe. They called the 1976 GMC motor home they traveled in Twinkie, Old Yeller, and Teddy, after Roosevelt, who signed the Antiquities Act in 1906, authorizing the president or Congress to set aside federal lands to protect significant natural, cultural, or scientific features.

The mayor of Lewisburg, John Manchester, welcomed us, talked about the economic benefit to the town of such a designation, and said he could see Lewisburg as a gateway city. Matt Kearns of the West Virginia Rivers Coalition also spoke. To bring awareness to the effort, he and a friend floated, hiked, and biked the headwaters of Laurel Run to the Elk, then the Greenbrier, the New, and the Kanawha into Charleston. They drank the water in the Elk but did not in the river near Charleston. On certain days two years ago, the people of Charleston could not drink out of taps when a chemical used in the coal industry spilled from a storage tank, contaminating the water supply.

At dinner before the show, I met a guy wearing an orange golf shirt who played a round at the nearby Greenbrier. He seemed like an important man about town as he knew the bartender by name, hugged one of the waitresses, and people came up to shake his hand, but he was not sure about the festival and said the Birthplace of Rivers was "controversial."

I brought that up with Mike Costello, director of the West Virginia Wilderness Coalition. He said people have found evidence that one monument does not allow hunting and therefore assume none will. "This one will allow hunting," he told me. Matt Kearns told me that people think you are shaking with one hand but have your fingers crossed behind your back. Rivers connect us, but they can be full of obstacles, as was this process. In a river, among the earliest forms of travel, you have to pick a way through hazards. Matt said that the only good argument against the monument was to keep people out, to keep it pristine.

Mike told me the monument would set aside a unique area on the East Coast. It would be more flexible than a national park, allowing mountain bikes for example and other forms of recreation. And they could continue the restoration of red spruce underway in the highlands. The trees once covered some five hundred thousand acres but were nearly wiped out by logging, now covered by hardwood. They cover about a tenth of the acres they once did.

In the morning, I wanted to see some of the proposed site. On my way up, I stopped at Beartown State Park, full of massive fragmented boulders, vertical cliffs, deep crevasses. "Hidden within the rocks is the story that really has no ending" said one of the signs on the boarded walk. White sand could be seen at the base of the cliffs, otherwise, all green fern and hemlock. A few rocks are scooped out enough that a person could climb in, nap. The horizontal strands I saw are related to deposits of sand and silt: "The ripple-like angular cross beds in the different rock strata were caused by current action when the rock-forming material was deposited along the shore of an ancient sea."

Walking around the boardwalk, I heard an energetic bird song I did not recognize. I recorded it and played it later for a naturalist friend. Clyde Kessler told me it was a winter wren, and "you have found a bit of holy ground when you hear that song." For Druids, the bird was sacred, and in England it is still known as the king of birds.

After Beartown, I stopped at McCoy's Market, a convenience store in Hillsboro, hometown of Pearl S. Buck, author of *The Good Earth*. Near the coffee pot are six guys in T-shirts and ball caps, big burly men. I wanted to talk to them about the monument proposal, but I realized, as I was in the bathroom, that I was wearing the Live Monumental hat given out as swag the night before. I didn't want to give myself away, nor did I want to get my ass kicked. So I took it off and went to the car to grab another. When I poured a coffee and joined the conversation, I got an earful about the government screwing everything up, taking away logging rights. Trust the government? Ask an Indian. The only good tree is a stump.

But the funny thing was that although these guys and I shared little in common, I enjoyed their company and laughter more so than with some of the folks I agreed with last night. Whatever their cause, they have not lost their sense of humor.

One of the biggest, Joe Walker, who drove a Neathawk Lumber truck, said why change? "It's perfect like it is."

For these men, coal and timber work are like farming in the way it becomes ingrained in the fabric of a community. Factor that in with a natural distrust of change. Accepting change, after all, would require admitting being wrong about something and that would require a kind of reflection and humiliation big burly men rarely engage in.

At that moment, a tall muscular guy they called "Doc" came in, "I'd sure like to operate on some people." He was retired from law enforcement for the Forest Service. They said this man, pointing to me, wanted to know what we think about the Birthplace of Rivers. "I think it sucks!" Doc said to a round of laughter. Besides, "the government screws up everything it touches," said the guy who worked for the government.

They told me more, about how timber sales helped with schools, about how the county is one hundred miles from one end to the other, a forty-five-minute drive, so transportation costs are high. They ran mills, heavy equipment, and one had a son that worked the limestone tumblers that keep the streams from being too acidic. But what caused them to be acidic? In part, acid rain. What causes that? The burning of coal. The mining of it too. Acid drainage from these mines also necessitates the treatment of the rivers.

They scoffed at the idea of tourism helping the economy. For proof, they asked me how much I spent on my coffee. Point taken.

I traveled on to the cranberry glades themselves, a forest in a bog. Some trees struggle to survive in the wet ground and acidic soil, but animals take up burrows in the hollow trees.

The glades are a great natural bowl in the nearby mountains, four thousand feet high, a misplaced tract of Arctic tundra in southern mountains, including reindeer moss. The namesake cranberry is an evergreen shrub more at home in the far north. Over ten thousand years ago, Ice Age glaciers pushed far south, and plants and animals suited for a colder climate took hold. In the area are several flesh-eating plants, sundew and purple pitcher, adapted to take in nutrients from insects because there are little nutrients in the damp habitat.

Red spruce would have blanketed the area but it was mostly clear cut. The slash the loggers left behind, treetops and limbs, created a tinderbox. And in 1930, Black Mountain burned to the ground, the heat so intense that the soil was consumed.

On both sides of the Highland Scenic Highway, I looked down on folds upon ridges, sunlight and shadow, cloud cover as varied in depth and shape as the mountains below. I was enjoying the scenery, as were others. I saw cars from New York, Ohio, Pennsylvania. At least some sign of tourism.

The area is defined by rivers, the Williams River to the north and the South Fork of the Cranberry River to the south. A former state black bear sanctuary and haven for animals such as bobcats, foxes, and the threatened northern flying squirrel, the Cranberry Wilderness is one of the wildest places east of the Mississippi. The West Virginia Department of Natural Resources lists the West Virginia northern flying squirrel, or "ginny," as highly vulnerable. There are fewer than one thousand "ginnies" left.

I ventured off the road to a trail, the North/South trail that runs east/west. Only a few hundred yards in, everything seemed unnervingly still. Even the light rain, what Scottish immigrants to the region would have called a smirr, seemed hushed.

Step by step, the path seemed to grow wilder until I was standing in waist-high ferns beneath dense stands of spruce. A few spring flowers held their bright color and the rhodie blossoms were about to burst among the many shades of green. There was the sweet smell of pinesap and another bird song I could not recognize but I had no recording device—I left the phone locked in the car. Besides, in nearby Greenbank is the National Radio Astronomy Observatory, which operates the world's largest radio telescope, and a radio frequency-free zone. The faintest signal can interfere with the telescope's detection of the music of the spheres. In the National Radio Quiet Zone, cell phones are restricted or banned.

I vowed to remember the song with some mnemonic I could look up later, some phrasing like "here Sam Peabody," Thoreau's for the white-throated sparrow, but I didn't write the *zee zoos* in my journal and the music was ultimately lost. I kept looking up at the sound. Cerulean warblers can be found in West Virginia, high in the canopy, "sky blue, sky high" say the books. Once plentiful in the United States, their population is decreasing faster than that of any other eastern songbird because much of their breeding habitat has been destroyed by mountaintop coal mining.

Whatever song it was, it left me in a kind of silent awe. No wonder birds like the winter wren inspire folklore.

I kept walking, and though my phone was in the car in some ways it was still with me, and with little effort my mind could churn through emails and to-do lists, putting me there and not here, but the river lay ahead. I trudged on, the trail growing less and less discernable, ducking under downed tree trunks and sloshing through ferns. I hiked on, stepping over coral-shaped fungi and meadows lush and majestic. Every few yards, there was a spider web that caught little but the morning's mist, tiny droplets like gems in a necklace.

Then I heard a bird I could recognize, the fluted *ee-oh-lay* of the wood thrush. Thoreau wrote that the bird "touches a depth in me which no other bird's song does," calling it "a Shakespeare among birds."

Jim McGraw, professor of biology at West Virginia University, has studied how they may be assisting wild ginseng in migration. Studying the birds on game cams and in the lab, he has found that the birds pick up the shiny red fruit but do not eat them. They spit the seed back out. But like the cerulean, their numbers are declining. That decline could affect ginseng populations that rely on thrushes for the free ride. And the plants may need to migrate as the climate warms. Ginseng is one of many plants adapted to a long-term temperature climate, so they will need to move to follow their optimal environment. Thrushes could be needed to help them get there.

In my bird reverie, I must have stopped paying attention to the cairns that marked the switchbacks because soon it felt like I was off the trail. Coincidently, at about the time I realized I was lost I became acutely aware of a loud crash of bushes below me. Possibly bear. How close to panic is Pan, god of woods and wild.

I bumped my head on branches and barked my shins on down to the North Fork of the Cranberry River, where I picked up the trail and river and camped for the night. I had one ear open for loud crashing noises and was awakened once by the split-crack-woomph of a downed tree, but I slept soundly, soothed by the water.

In the morning, I hiked back to the car and stopped in Marlinton at the Dirtbean for coffee, where I ran into Matt Kearns again. He was meeting with several other people, including the CEO of the Snowshoe ski resort, Frank DeBerry, to talk about the recreation economy. It was a good place to have these discussions: at the back of the Dirtbean, they rent bikes for the Greenbrier Trail. Fat tires and a coffee to go.

At the moment, much of the county's economy comes from timber sales. But could lodging taxes provide a new revenue stream? Currently, 85 percent of tourists come in winter for skiing. Matt and others talked about adding spring, summer, and fall. He saw the Birthplace of Rivers idea as part of a climate change survival plan, as a way to take up some of the slack and hedge against a decrease in the warming that could affect winter tourism. DeBerry wanted to offer more seasonal activities as an offset.

Many people were looking for a silver bullet to bring jobs back again. From Matt's perspective, there were good paying jobs but not as many, "so we have to adapt. Instead of one company that brings a thousand jobs, maybe there are a hundred that bring ten."

I mentioned the fear and uncertainty I heard in the market of Hillsboro, their "why change?" sentiment. He said they were doing their best to provide information. "If you like the way the area is currently being managed, we will keep that plan, but lock it down the way it is. So it doesn't change. It's the change, new kinds of extraction or drilling, we're helping to guard against." Matt said they have to work very hard to keep the same or unwanted change will happen. For Kearns, protecting public lands could be like protecting the family jewels.

I said the guys in the market thought the idea was coming from outsiders. "I chafe at the notion that I'm not from around here." He showed me the tattoo on his arm, the state motto of West Virginia, *montani semper liberi*, mountaineers are always free.

Several days later, back to routine and plugged in, the meaning of temporarily getting lost in the wilderness started to become clearer. I thought of writer Wallace Stegner's well-known "Wilderness Letter," that "without any remaining wilderness we are committed wholly, without chance for even momentary reflection and rest, to a headlong drive into our technological termite-life."

Less well-known is the inspiration for his devotion to wilderness. In "Overture," he said camping by a river, the Henrys Fork of the Snake, was what made him give himself to wilderness. Among the spray-cooled and scented spruce, a "symphony of sound," "I gave my heart to the mountains the minute I stood beside this river with its spray in my face and watched it thunder into foam."

By the Cranberry, I spied a few brook trout disappearing into the shadows. They predate the glaciers and are a sentinel species, threatened by warming waters. However, Than Hitt, a fish biologist with the U.S. Geological Survey, said they could be more affected by flow. He saw some adaptations to heat, some thermal tolerance, but worried that the increased water and storms would scour the streams during their breeding season, washing away eggs and laying habitat.

Camping beside the river, I drank in the music, hiss and splash in the main channel, lower murmur on sides, some faint gurgle in the back and beneath it all pulsing a steady rhythm and flow, changing yet everlasting.

I went back up to West Virginia to cross-country ski the last weekend in January 2017, one of the first there had been enough snow, in a winter likely to be the warmest yet, each warmer than the one before. I was up there seeking some solace from the week's news of building walls and banning people. On the road into Whitegrass near the Dolly Sods Wilderness in West Virginia, they fly flags of many nations attached to alder birch saplings, welcoming everyone.

The cover was thin, a little sticky, and I found more than a few roots and rocks but also more and more snow as I went higher, as if traveling to some mythical land where it would be higher and deeper yet. At one spot, by a small creek and waterfall, I took in the stillness of waters, even among the rushing sound of the creek and the dripping icicles. I thought of the springs that fed it, the layers of rock cradling it, the forest surrounding it, and the far ridges beyond which I still had time in the day to explore. Leaning into my poles, catching my breath, as I thought of my own past trips and of natural processes, a deep sense of time began to form, which also made it possible to imagine a deep future.

Paying attention to the hardwoods on the way up, the greenish mosses on the north side of the bark, of the movements and tracks of animals, invisible neighbors, lightened my spirits after a fear-filled and disappointing week. The only tweets were high up in the canopy above.

After the morning's ski and into late afternoon, legs rubbery, I skied down to the bottom near a bubbling sand springs, artesian pressure pushing up from below. A family with young kids was there, taking a well-deserved snack break, hypnotized by the waters. The kids were asking where it came from, where it went. Their curiosity fed their imagination, which nourished overall well-being and wonder. The water would join with that creek I stood near, would travel through the boggy glen of the Canaan Valley Wildlife Refuge, on to the dark forested Blackwater north and west and eventually on down to the Mississippi valley where it would join with the mountain waters of Stegner's West.

The Birthplace of the Rivers National Monument never happened under President Obama and its future looks dim under a Trump administration. It had grassroots support but never garnered enough from local or state officials. Matt Kearns of West Virginia Rivers Coalition told me they hand delivered over two thousand emails, postcards, and letters to the White House, but "without the buy-in from our congressional delegation, Obama wasn't willing to wade into our battle."

To make matters worse, a week after the film festival in Lewisburg, the region experienced the third-worst flood in state history, with the worst one occurring in 1972—a rainfall so hard that a dam built for a coal slurry pond dislodged and ravaged the community of Buffalo Creek. The rain caused the level in the pond to rise and burst through, spilling sludge on the community below. An enormous wall of black waste swept through the hollow, killing 125 people and destroying "everything in its path," the title a 1976 book about the disaster.

Visiting the area a year later, sociologist and author Kai Erikson found the people traumatized. Many relived the events in flashbacks. In his book, Erikson documents a collective trauma, not just from the physical objects carried away but from the social norms, institutions, and traditions carried on for generations. Without them, it was difficult for residents to find meaning and purpose. Nearly everyone he met suffered from some combination of "anxiety, depression, insomnia, apathy, or simple 'bad nerves.'" They grasped for an identity when the usual bases for it were no longer present, and they lost their sense of community. The experience was repeated during Katrina in New Orleans. Ron Eyerman, in the title to his book about cultural trauma and Katrina, asks *Is This America?* As Robert D. Bullard and Beverly Wright say, "days of hurt become years of grief, dislocation, and displacement." According to the literature on trauma I have read, the experience goes something like this: the past is awful, the present is terrible, and the future looks bleak.

Climate change could take a significant toll on mental health. The American Psychological Association and ecoAmerica, a nonprofit that works for climate solutions, released a report in 2017 called *Mental Health and Our Changing Climate: Impacts, Implications, and Guidance*. It builds on earlier work that addresses ways to lift people up out of the grief and trauma associated with the stressors generated by climate change.

After the current flood, the Greenbrier golf course was littered with severed trees, buried under mud. Though hardly traumatic, they had to cancel their Greenbrier Classic PGA. More than two hundred homeless people were invited to bunk at the five-star hotel. Grills were put out to cook food. People donated water and perishables. Catastrophic, traumatic times can also bring out the best in people, define us, remind us of our common lot.

According to Kevin Trenberth, a senior scientist at the National Center for Atmospheric Research, there is about 4 percent more moisture in the atmosphere since 1970. West Virginia and the northeast have experienced a 71 percent increase in extreme rainfall events since 1958, says the Third U.S. Climate Assessment, a problem particularly acute for the steep-sloped mountain state with miles of run-off. It's difficult news for a state whose economy has been built by a fuel responsible for a quarter of all greenhouse gas emissions.

In the lodge at Whitegrass, which always smells like wood smoke and soup being made, where the staff all sport wet ski boots, I asked owner Chip Chase how he was faring in the warm winter. He makes no snow but "snow farms" by putting up fencing to catch drifts, and his wife, Laurie, runs a natural food café. People eat bowls of chili and chocolate chip cookies around the pot-bellied stove. He was hopeful winter would finally arrive, but "we've built a resilient little business here, so don't worry about us." In other words, they have diversified their economy.

On my ski to the very top of Weiss Knob, I could see Mount Porte Crayon, named for a writer-illustrator (also known as David Hunter Strother) who published often in *Harper's Monthly* in the mid-nineteenth century and loved the nearby highlands. In his *Virginia Illustrated* (1857), he writes of his visits to Canaan and the region's many springs, helping to popularize them. At the close of the book, he writes of hidden deep valleys overshadowed by gloomy hemlocks and a pervading sense of toil, even "oppressive loneliness." But occasionally, a brook appears, and then "a hoarse murmuring deep down in the earth." Now a feeling of awe and mystery "steals over the spirit." A little farther, he sees a "broad, bold river burst suddenly, its crystal waters flashing in the sunshine, roaring and leaping," "rejoicing the lonely valley with the voice of music, and the eye of the wanderer with the freshness of beauty." In some human hearts, he writes, "the course of love is like that mountain stream."

Though it never achieved monument status, we can still enjoy the wild and wonderful in West Virginia for now, when we need wilderness more than ever. There are other jewels in these mountains than the ones underground, timeless ones like warbling birds and spring-fed streams. "Knowing wilderness is there," wrote Sen. Clinton Anderson in 1963, "we can also know that we are still a rich nation, tending our resources as we should—not people in despair searching every last nook and cranny of our land for a board of lumber, a barrel of oil, a blade of grass, or a tank of water."

Wallace Stegner said the idea of wilderness alone could sustain us, but we need the real thing too, the actual contact with it. Up there in those headwaters, I found the source of so much, rowdy creeks running wild and free, hot springs and cold ones, a symphony of nature and a source of rivers, a bubbling richness. If not a geography of hope, to use Stegner's phrase, it could certainly be the birthplace of it.

9

More Ghosts on the Coasts, and the Last Place to Go

As I traveled to talk to people and communities facing sea level rise, I have seen them. I have heard of living shorelines, green infrastructure, raised roads, and scuppers on storm drains to let water out but not in. I encountered alligators and sea turtles, ancient species, mostly unaffected, for now. After Hurricane Matthew, I rescued a fish from a puddle in Charleston, swept in over the seawall, now returned to the sea. Sunny day flooding in Norfolk, beach erosion in Nag's Head, remnants of vanished cultures on the Georgia isles, one still clinging in Louisiana, but what I am most struck by are the trees.

In my mind's eye, they can still be conjured, gray and weathered, spectral, wraithlike. "Boneyard beaches" is one term I have heard for stands like these on the coast; "ghost forests" is another. The skeletons stand steadfastly upright in the sand, pretending to be the trees they once were. Their denuded forms give an odd aspect to the beach. This is no place to throw down a towel. No place to lather your body in lotion. Instead of beach blanket bingo, the atmosphere is funereal.

The trees were killed where they stood by the constantly eroding and unsheltered shore, by the salt air blasting them, the soil degrading into silicate, the briny water drowning roots, the tide and sea rising, rising. Bleached by sun, they look like giant claws, reaching up out of some villainous ocean or marooned vessel.

Humans could move further inland. Pack up and leave. But trees can't run from impending doom, which is perhaps why they make such enduring symbols in literature. Think of the oak tree outside of Boo Radley's house, repository for small gifts, symbol of friendship, thriving through adversity. Or that tree that grows in Brooklyn, invasive ailanthus, sprouting upward as if parallel to young Francie Nolan's impoverished circumstances.

But these ghost trees seem to signify no such hopefulness. If they are a tree representing knowledge, some sinfully delicious fruit, they are a portent of things to come, the ghosts of climate change.

Of course, dead trees, in every stage of decay, are vital to the living. So we might not mourn their passing. Birds, bats, and raccoons live in the crevices and hallows. Squirrels climb inside for shelter and to store food. Insects eat the moldering wood. From death comes life, right?

Once infected with salinity, these ghost forests fail to produce new saplings. A healthy forest is like a family, where plants and animals with shared interests coexist. They help each other become safe, secure, and strong. The tall trees are the adults, providing shade and shelter for the younger ones. Pines, sweet gums, and maples grow in these forests, but the granddaddies are the live oaks and cypress. They live for hundreds of years, and the Angel Oak at John's Island in South Carolina is believed to be anywhere from 500 to 1,500 years old.

The Angel Oak is a southern live oak, a tree native to the Lowcountry. When I visited it, I was stunned at the sheer massiveness of the base, more than twenty-five feet with some seventeen thousand square feet of canopy cover under its branches. I crept around under the spidery limbs, my imagination wandering. In *A Southern Journey: A Return to the Civil Rights Movement* (1997), Charleston resident Bill Saunders told poet Tom Dent the tree is said to be a symbol of survival, bending but not breaking after storms. "When the sun comes out again after the storm, the Angel Oak is still there, standing majestic no matter how many limbs were taken."

The oak derives its name from Justus Angel, a slave owner. Local folklore tells stories of ghosts of former slaves appearing as angels around the tree. It is believed that some were lynched there, the site of a former plantation. There is an indescribable energy to trees this big, as if spirits surround them.

In *Ghosts and Legends of Charleston* (2010), Denise Roffe interviewed a woman descended from the slaves who toiled on the island's plantations. She recounted the legends of the tree, including that it was once home to huge birds (likely vultures), who fed on the bodies of lynched slaves. The old woman continued saying that many people were buried under the tree including Native Americans who met under its shady branches. She stated that these spirits still gather around the oak and that they also work to protect the tree.

The dead stay with us, this much we know. Our ghosts reflect the contours of our anxieties, the nature of our fears, and possibly our hopes for return. Ghosts haunt us because we want to believe in an afterlife, for death to mean something. For what they tell us about changes to the coast, tree ghosts should make us shudder.

We also know that trees communicate. They talk through underground mycelial networks. Threatened with a parasite, trees can turn bitter. And they can withhold nuts until animal populations are reduced. They know how much soil is available to them and where to stretch their roots. Trees, like animals, can adapt. But not fast enough to keep pace with a rising sea. And while we suspect that animals can sense grief, can trees mourn the dead?

If they can, they are full of it right now, as are many inhabitants of the coast. Among the many adaptations we may have to make as the sea rises, in addition to the physical ones, is coping with and recognizing our grief. Grieving brings people together, to remember and reflect yes but also to forge ahead and solve problems. In evolutionary terms, grief for a missing partner led to a search and a reunification, living to see another day, where resources and nurturing would once again be shared. The precursor of grief is love, the positive emotions experienced even before grief comes into play. That adaptive trait is basically an understanding of how our mutual dependence and cooperation increases our chance for survival.

In my lifetime, I have lost some important trees. The ash outside my bedroom window that I would stare up into as a boy—emerald ash borer—gone. The hemlocks at the first home I purchased, shading the back deck and swing-set—wooly adelgid—dying. Because of these losses I am tempted, on walks when feeling particularly spry, to want to thank the trees. For the shade they provide. The oxygen they give. The soil they hold in. But that would be ridiculous, right? Thank a partner for the good work they do? The love for trees, and the corresponding grief for those lost, taps into this feeling of kinship: their demise could signal something ominous for us.

On Cumberland Island National Seashore, at the Carolina coast, I get close to probably a live oak, perhaps a pine, reach out a cautious hand, as if touching a body laid to rest. The smooth, hard texture touches back somewhere under the skin. I look up into the bare canopy, still spectacular in shape and structure, and imagine it as a sapling, maybe a hundred years ago or more, on a much different shore. To defeat these ghosts, and the larger problem they signify, you must first understand how they are created. You must be able to decipher between what is natural and what is man-made, to tease out hoax from truth, and be willing to wade in the muck and mire where these ghosts linger.

We walk in a healthy woods to smell the spring rain, look for the burst of wild color, bath in cool air and damp shade. But all along the coast, and places further inland decimated by drought or bark beetle or both, these dead trees show symptoms of a larger planetary illness. They intrude on our notions of the pastoral, telling a macabre story of a species who fails to see the forest for the trees.

As a species still evolving, we are a fascinating mix of art and animal. We create amazing structures, paintings, and scientific breakthroughs, yet we have also, through our impulses, sown the seeds of our own destruction. We now stand at the threshold of something that has never existed before: the knowledge that we could be robbing our children of a future. We take more than our share of space, but we are now also curtailing time, over-borrowing on available carbon, when the heirs of our debt had little say in the terms.

Maybe we will pull through and get this right. I have talked to many who think so. I keep wishing it too, though my thoughts ebb and flow as if with the tide. Maybe someday this particular moment will be remembered as some historical footnote, like a plague or war, and our polluting, wasteful ways will be looked upon as associated with some former Dark Age, just a blip in the line graph of carbon parts per million, a swelling in the ice core before we finally got our ark together. Every former civilization has some kind of flood myth in its literature. Terence J. Hughes, a retired University of Maine glaciologist, has said that these stories are more than fairy tales. "Some kind of major flood happened all over the world, and it left an indelible imprint on the collective memory of mankind that got preserved in these stories."

Sometimes it gets cold out and snow blankets everything, absorbing stress and sound, and I think it will all be okay. Or I will be amazed at what

I will see by taking a walk. Just the other day, I saw a bear on my bike ride. We heard but did not see a bobwhite on our farm. Nature is trying to pull through, fighting for survival. It knows nothing else. But so much is missing from the frame.

The planet is changing because we changed it. We could still ourselves change and allow things to come back to some sense of what they were, or we can keep hurtling toward the cliff (or the sea will just rise to meet us where we are). But it's no small task to bring back what has been lost: the coral reefs, the drought-stricken soils, the degraded forests, the ocean's chemistry.

If we can recognize our role in creating the change, we can come around to our part in the healing, not just as individuals but as citizens of a much larger community. The same thing that will help our long-term needs, an appreciation of our shared dependence, will also take care of many of our short-term problems. We have to move past the notion that we can go it alone. No island can right now.

If there is a silver lining in pulling out of the Paris Climate Agreement, perhaps it is that cities, states, and local communities will be even more committed. They are already widening their networks, like expanding ripples in water, cooperating with one another. New Orleans is talking to Norfolk. Miami shares information with other cities. However, without a coherent national strategy, they are resilient isles in a threatening sea.

Some have said that attention to climate change is just another way to make money, but the scientists I have talked to scoff at this. They are making little money in the pursuit of knowledge. But innovations are happening, those that mimic the power of natural systems, harnessing wind and sun, without the waste and without setting the world on fire.

At the moment it seems like we are burning down the house to stay warm. We are surely capable of enriching our planet rather than diminishing it, as we would our own soil or property. The practicalities of our situation, preventing ourselves from drowning or burning, are not only scientific, they are moral.

When I have visited coastal places, I am struck by how much is changing, how much could go wrong, but also, in the cases where the island was once under water under a previous epoch, or formed during a more recent one, that sand is a kind of stardust, and beneath the sinking ground is the very energy we would tap to sustain us, an energy that also shoots through us and every living thing in the form of sunlight. The sun seems to give not only physical energy but a metaphysical one too. In such moments, it's hard not to feel a connection to all, including to poets of the past like Whitman, "throwing myself on the sand, confronting the waves," chanting pain and joy, "uniting here and hereafter," time and place.

Whitman had a tremendous capacity for imagination and feeling, and it sometimes seems we are mentally imprisoned, unable to soar in our minds to see our actions from high above, like in a satellite image, or the amazing picture of earthrise taken from Apollo 8 during Christmas week, or from the long view of history.

As much as climate change is a geophysical or climatological concept, it is also a very social one, with its own vocabulary, tradecraft, and protocols. Traveling through the South, I sometimes felt like a spy in enemy territory, there to uncover some secret black-ops. I worried about approaching some people, but it always seemed we could have a conversation. I agree with Ed Maibach of George Mason's Center for Climate Change Communication who told me you can always talk to people, even in the reddest of counties. "It may not be a top tier issue, and they may have some facts wrong, but you can have a conversation."

Hope may be our ultimate adaptation, filling in the blanks and uncertainty. If our present condition were a puzzle, hope may be the missing piece, available to all people in all situations.

When Thoreau's two-hundredth birthday recently came around (July 12, 2017), some reflected on how far we had come. Thoreau saw nature differently from many of his time. Most then saw it as inexhaustible, God-given, there for the taking, something we could keep acting upon forever. On beaches with so much wind and water, I have sensed this very same vastness. How could something so big be harmed? But in the past two hundred years, we have come around to viewing nature as something very finite and delicate, and that we are witnessing a particular moment in time with respect to the nonhuman world: that the planet is vulnerable—an important shift. Thoreau worried that he would come to the end of his life and fear he had not lived. As bad, or worse: to come to the end of our lives, at this particular moment in history, and when the time came to do something, we did not even try.

When will the momentum shift? Little moments of incremental progress, big steps back, but still the number of people understanding climate change as a threat is growing. Can we make it? Is there enough time, and are we smart enough to catch up to the change we have ourselves created?

When traveling to southern places, I often brought along reading material associated with them. In Georgia, I reread Flannery O'Connor's "A Good Man Is Hard to Find." On the family's way out of town, the son calls Tennessee a "hillbilly dumping ground." "Georgia's a lousy state too," he says. The grandmother is full of piety and old traditions, tells the children not to talk that way. She brings up the "Misfit," the killer, because she does not want to see him, but she does. When he shoots her son, Bailey, the Misfit comes back wearing Bailey boy's shirt. Just as the grandmother reaches out, "you're one of my babies. You're one of my own children," the Misfit shoots her, saying she "would have been a good woman if it had been someone to shoot her every minute of her life." The point is that we accept grace too late. We act the way we should act only when under the gun. We are under it.

Creatures great and small will bear a heavy burden of our ignorance, selfishness, and stupidity. I take some solace in knowing that they will always do everything they can to stay alive, no matter what we have done. They and their many adaptations are an important guide.

So much still survives. Summer travelers today often head to the beach. I often head in the other direction to the public lands of the high mountain forests in the heat of summer. Fashionable resorts were built near there in Asheville, North Carolina, or White Sulphur Springs, West Virginia. The air is cooler there, the sun shaded. In the high country, speckled brook trout somehow survive under tumble down streams, are somehow miraculously even there in the first place. Lungless salamanders scatter under mossy rooks, disappear under leaves or tunnels in the soil. A long time ago, things were wetter and more suitable for these species. But as things changed and glaciers withdrew, they retreated to higher and higher habitats that were more suitable—the last places to go.

Their journey may presage a retreat for more modern human refugees. The emptied-out towns in the Appalachian coalfields could once again be viable as people seek cooler and higher places to live.

If we do flame out, we will share a fate with dinosaurs. But they had no knowledge of their doom. Our brains may be our greatest evolutionary advantage, but they make our self-destructiveness all the more inexcusable.

Come to think of it, I'll head to the high country in winter too, looking for snow, or in fall, to see the colors change, and in spring, looking for the wildflowers, increased birdsong, the renewed promise of life. Everything is on the move then, a grand coming-out-of-the-ground party. We made it through the winter, in whatever form it took. I'll send out a wish that we use those big, adapted brains for good, that we make wise decisions when we might have done something else, affirming what is best for both us and nature. For hope, I'll look to the tall trees above, now coming back in bloom, filling with voices, attuned to the moment, looking for life, a beginning.

Acknowledgments

To write this book I had a lot of ground to cover, stories that unfolded both horizontally across the South and vertically in the changes through time. To learn about the latter, I am especially grateful to the many experts who shared their time and knowledge. A complete list can be found in the end pages, but many went above and beyond the call of duty. My extended thanks to Chris Moore and Thomas Quattlebaum of the Chesapeake Bay Foundation for a day on the *Bay Oyster* and Carol Pruitt-Moore of Tangier Island for a ride in *To Oz*. In North Carolina, thanks to Mike Bryant, Dennis Stewart, and Chris Oishi. In South Carolina, Carolee Williams and Laura Cabiness answered many of my questions in person or by email. In Georgia, thanks to Crawfish Crawford, to Paul Wolff for the deck to sit on, and Carol Ruckdeschel for her picnic table and tour (and for towing some of our gear). Thanks to Jason Evans for walking around the Indian River, Hal Wanless (who I owe a lunch), Chris Bergh for the tour of Big Pine Key, and Steve Traxler for the hike up Centennial Trail. Thanks to Mark Davis of Tulane, Ricky Boyett and John Lopez in New Orleans, as well as John Anderson and Jim Blackburn of Rice University and Bill Merrell of Texas A&M. In West Virginia, I enjoyed a most pleasant evening with Lou Dawson Stoker. Thanks to Dana Stoker Cochran for setting it up. Thanks also to Mike Costello, Matt Kearns, and Ben Gilbert for talking with me. Thanks also to the many people, named and unnamed, who spoke with me on the street, in the bar, by the water. Once again, please pardon the intrusion. I hope I have represented your communities both faithfully and respectfully.

To get to these places, I could have used a full-time travel agent. Holly King helped with many arrangements. Were I a NASCAR driver, I would have to thank the good folks at the local Enterprise. I may have driven more miles than they bargained for. Thanks to Rosemary Guruswamy and Kate Hawkins of Radford University for helping to make these trips possible.

A number of people put me up (put up with me). Thanks to Don Samson, Pat Lackey, and Diana Dul. Also, thanks to Tim and Elene West for their continued support and generosity.

Portions of these chapters appeared elsewhere. Thanks to Tom Lynch, O. Alan Weltzein, Susan N. Maher, and Drucilla Wall for their work with "The Proximity of Far Away: Climate Change Comes to the Alligator" in *Thinking Continental: Writing the Planet One Place at a Time*; to Kathryn Kirkpatrick for "Ghosts on the Coast," *Cold Mountain Review*; Jessica Cory for "Take in the Waters," *Mountains Piled upon Mountains: New Appalachian Nature Writing*; and Hattie Fletcher and the crew at *Creative Nonfiction* for "Ebb-tide Optimism on a Climate-Changed Coast."

I thank Tim Poland and Dan Woods for their help both on the page and the stream (and on the porch, when Tim remembers to reserve it). Good to know they can be plied with Scotch. Both John Lane and Janisse Ray turned up a number of leads and resources, as did colleagues Tom Hallock, Tanya Corbin, Carter Turner, and Bill Kovarik. Scott Slovic provided valuable insight into early drafts and the finished product. Susan Harris improved clarity and consistency, and it was a pleasure to once again work the University of Georgia Press.

Sam Van Noy has a career as a long-haul trucker if he wasn't so good at finding wildlife. Elliot Van Noy shares a page with Ralph Waldo Emerson. A very good start to her writing career. Catherine Van Noy, eagle-eyed reader, owl-eared listener, also came with me on some of these trips and shares the big, long journey with me.

Notes

Chapter 1. Tombstones by the Sea: Our Climate Change Commitment

Page 2. **Since the early 1900s, about two-thirds of species studied** Camille Parmesan, "Ecological and Evolutionary Responses to Recent Climate Change," *Annual Review of Ecology, Evolution, and Systematics*, 2006.

Page 2. **first leaf growth of honeysuckle and lilacs across the lower forty-eight states** Andrew J. Allstadt et al., "Spring Plant Phenology and False Springs in the Conterminous US During the 21st Century," *Environmental Research Letters*, October 2015.

Page 2. **their ranges were moving** Camille Parmesan and Gary Yohe, "A Globally Coherent Fingerprint of Climate Change Impacts Across Natural Systems," *Nature*, 2003, 37–42.

Page 3. **researches have found that marine plants and animals** Elvira S. Poloczanska et al., "Global Imprint of Climate Change on Marine Life," *Nature Climate Change*, 2013.

Page 3. **global temperatures rise** NASA, "Long-term Warming Trend Continued in 2017," January 18, 2017, https://climate.nasa.gov/news/2671/long-term-warming-trend-continued-in-2017-nasa-noaa.

Page 3. **climate change is largely irreversible** IPCC, Fifth Assessment, "Long-term Climate Change: Projections, Commitments and Irreversibility," 2013, 1033.

Page 5. **carbon dioxide climbed to over four hundred parts-per-million** NASA, "Carbon Dioxide," https://climate.nasa.gov/vital-signs/carbon-dioxide.

Page 6. **No one under thirty** Richard Rood, "30 Years of Above-average Temperatures Means the Climate Has Changed," February 26, 2016, https://phys.org/news/2015-02-years-above-average-temperatures-climate.html.

Page 6. **1.1 millimeters per year before 1990** Sönke Dangendorf et al., "Reassessment of 20th Century Global Mean Sea Level Rise," June 6, 2017, Proceedings of the National Academy of Sciences, www.pnas.org /content/114/23/5946.

Page 6. **rate of sea level rise at 3.4 millimeters** NASA, "Sea Level," https://climate .nasa.gov/vital-signs/sea-level/.

Page 6. **we basically have three choices** James Kanter and Andrew C. Revkin, "World Scientists Near Consensus on Warming," *New York Times*, January 30, 2007.

Page 7. **stories of the dream, not the nightmares** Bill Kilby, "A Psychologist Explains Why People Don't Give a Shit about Climate Change," *Vice*, June 9, 2015.

Page 8. **economic harm that could afflict southern places** Solomon Hsiang et al., "Estimating Economic Damage from Climate Change in the United States," *Science*, June 30, 2017.

Chapter 2. Our Best Defense: Shelling the Naval Base, Virginia

Page 14. **thirty-five million years ago when a fiery meteor** USGS, "The Chesapeake Bay Bolide Impact: A New View of Coastal Plain Evolution," https://pubs.usgs.gov/fs/fs49-98/.

Page 16. **tidal gauge at Sewell's Point** NOAA, Tides and Currents, "Mean Sea Level Trend, 8638610, Sewell's Point, Virginia," https://tidesandcurrents.noaa.gov /sltrends/.

Page 17. **Three-quarters of Virginians live** Steve Nash, *Virginia Climate Fever* (Charlottesville: University of Virginia Press, 2014), 72.

Page 21. **Norfolk is experiencing more like 4.5.** NOAA, Tides and Currents, "Mean Sea Level Trend."

Page 22. **rebounded thirty-two millimeters** Sustainable Water Initiative for Tomorrow, "Fact Sheet: Land Subsidence Eastern Virginia's Increasing Vulnerability to Sea Level Rise," http://swiftva.com/wp-content/uploads/2017 /03/LandSubsidence.pdf.

Page 23. **Schulte authored a 2015 paper** David Schulte et al., "Climate Change and the Evolution and Fate of the Tangier Islands of Chesapeake Bay, USA," *Scientific Reports*, 2015.

Page 28. ***New York Times* article about the fate of Tangier** Jon Gertner, "Should the United States Save Tangier Island from Oblivion?," *New York Times*, July 6, 2016.

Page 30. **Taylor told a radio reporter** Taylor in Matt Laslo, "Not All GOP Members Deny Climate Change. But They're Still the Minority," April 18, 2017, wvtf.org/post/not-all-gop-members-deny-climate-change-theyre-still-minority.

Page 30. **climate change contrarian** Michael Mann, *The Hockey Stick and the Climate Wars: Dispatches from the Front Lines, (New York: Columbia University Press, 2012), 106.*

Page 36. **the program is $25 billion in the red** Government Accountability Office, "National Flood Insurance," www.gao.gov/highrisk/national_flood_insurance /why_did_study.

Page 36. **Those companies made profits** Laura Sullivan, "Business of Disaster: Insurance Firms Profited $400 Million after Sandy," May 24, 2016, www.npr .org/2016/05/24/478868270/business-of-disaster-insurance-firms-profited-400 -million-after-sandy.

Page 36. **$100 billion of coastal property could be underwater** Risky Business, "National Report: The Economic Risks of Climate Change in the United States," 2016, https://riskybusiness.org/report/national/.

Page 36. **designated high-risk zones** FEMA, "Basic Facts about the National Flood Insurance Program," September 19, 2016, www.fema.gov/disaster/4277/updates /basic-facts-about-national-flood-insurance-program.

Page 37. **During the Ice Age, oysters increased** Roger Mann et al., "Reconstructing Pre-colonial Oyster Demographics in the Chesapeake Bay, USA," *Estuarine, Coastal and Shelf Science*, 2009, 221.

Page 38. **fifteen million bushels a year** Fred Powledge, "Chesapeake Bay Restoration: A Model of What?," *BioScience*, December 1, 2005.

Page 39. **John Smith, for all his swagger and bluster** Francis Jennings, *The Invasion of America: Indians, Colonialism, and the Cant of Conquest* (Chapel Hill: University of North Carolina Press, 1975), 77.

Page 39. **The mildness of the air** John Smith, "From the General History of Virginia," in *Reading the Roots: American Nature Writing before Walden*, ed. Michael P. Branch (Athens: University of Georgia Press, 2004), 56.

Page 40. **Nature Conservancy have created maps** Nash, *Virginia Climate Fever*, 72.

Page 40. **die from extreme heat exposure** CDC, Press Release, July 10, 2001, www.cdc.gov/media/pressrel/r010713.htm.

Page 40. **community of living beings, a collection of species** Lynn K. Nyhart, *Modern Nature: The Rise of the Biological Perspective in Germany* (Chicago: University of Chicago Press, 2009), 152.

Chapter 3. The Proximity of Far Away: Climate Change Comes to the Alligator, North Carolina

Page 44. **sufficient catalyst** Patrick D. Murphy, "Pessimism, Optimism, Human Inertia, and Anthropogenic Climate Change," *ISLE: Interdisciplinary Studies in Literature and Environment*, 2014.

Page 45. **90 percent of the heat generated** Intergovernmental Panel on Climate Change (IPCC), Fifth Assessment, "Summary for Policy Makers," 2014, www.ipcc.ch/report/ar5/.

Page 45. **70 percent of Americans** Anthony Leiserowitz et al., "Climate Change in the American Mind," Yale Program on Climate Change Communication, 2017, http://climatecommunication.yale.edu/publications/climate-change-american-mind-march-2016/.

Page 49. **North Carolina loses about three feet of shoreline per year** N.C. Division of Coastal Management, Policy and Planning Section, "Oceanfront Construction Setback Factors," 2014, http://portal.ncdenr.org/web/cm/oceanfront-construction-setback.

Page 49. **beating our head against the wall** Sara Peach, "Rising Seas: Will the Outer Banks Survive?," *National Geographic*, July 24, 2014.

Page 49. **string of pearls** Stanley Riggs et al., *The Battle for North Carolina's Coast: Evolutionary History, Present Crisis, and Vision for the Future* (Chapel Hill: University of North Carolina Press, 2011), 104.

Page 50. **The main problem they have is fear** Lori Montgomery, "On N.C.'s Outer Banks, Scary Climate-Change Predictions Prompt a Change of Forecast," *Washington Post*, June 24, 2014.

Page 54. **sometimes exceeds seventeen Foot long** John Lawson, "From *A New Voyage to Carolina*," in *Reading the Roots: American Nature Writing before Walden*, ed. Michael P. Branch (Athens: University of Georgia Press, 2004), 107.

Page 54. **senses into such a tumult** William Bartram, "From *Travels through North and South Carolina, Georgia, East and West Florida . . . 1791*," in Branch, *Reading the Roots*, 108–9

Page 56. **tree ring analysis on bald cypresses shows** David W. Stahle et al., "The Lost Colony and Jamestown Droughts," *Science*, April 1998, 564–67.

Chapter 4. Fish out of Water: High Tide in the Lowcountry, South Carolina

Page 61. **significant tidal flooding events per year** City of Charleston, *Sea Level Rise Strategy*, December 21. 2015, http://www.charleston-sc.gov/DocumentCenter/View/10089.

Page 65. **megacities are situated in coastal areas**. Pope John Francis, "Laudato Si," https://laudatosi.com/watch.

Page 66. **For the past twenty years** Mark Sanford, "A Conservative Conservationist?," *Washington Post*, February 23, 2007.

Page 67. **help fellow conservatives to see in climate change** Bob Inglis, in "Director of Mason's Energy and Enterprise Initiative Receives JFK Profile in Courage Award," May 1, 2015, https://www2.gmu.edu/news/1434.

Page 67. **counter the socialism-inspired Young Pioneer Camps** American Legion, "About Boys State," www.legion.org/boysnation/stateabout.

Page 68. **to introduce doubt where very little exists** "Bob Inglis: Climate Change and the Republican Party," *PBS Frontline*, October 23, 2012, www.pbs.org/wgbh/frontline/article/bob-inglis-climate-change-and-the-republican-party/.

Page 68. **having a relationship with god** "Katharine Hayhoe: Evangelical Christian, Climate Scientist," *Biologos*, November 9, 2012, https://biologos.org/.

Page 68. **Inasmuch as it hasn't been proven** "Bob Inglis," *PBS Frontline*.

Page 69. **foil to anxiety and depression**. In Beth Azar, "A Reason to Believe," American Psychological Association, December 2010, www.apa.org/monitor/2010/12/believe.aspx.

Page 70. **biospheric egalitarianism** Katharine Wilkinson, *Between God and Green: How Evangelicals Are Cultivating a Middle Ground on Climate Change*, (New York: Oxford University Press, 2012), 91–92.

Page 70. **why care about climate change** Bill Moyers, in Wilkinson, *Between God and Green*, 91.

Page 70. **engagement in relief and development work** Wilkinson, *Between God and Green*, 127.

Page 71. **Where there used to be sand dunes is now rock** "Matthew Flattens, Scarps Lowcountry Beach Dunes," *Post and Courier*, October 7, 2016.

Page 73. **headed for the Sea Islands that slid down** Toni Morrison, *Beloved* (New York: Penguin, 1987), 131.

Chapter 5. Ebb-tide Optimism: Ghosts on the Golden Isles, Georgia

Page 85. **the sea level is rising** Justin Gillis, "Flooding of Coast, Caused by Global Warming, Has Already Begun," *New York Times*, September 3, 2016.

Page 85. **I'm breathing a sigh of relief from what I saw** Rae Ellen Bichell, "Hurricane Matthew Causes Floods on the Low-Lying Barrier Tybee Island in Georgia," October 9, 2016, www.npr.org/2016/10/09/497256647/hurricane-matthew-causes-floods-on-the-low-lying-barrier-tybee-island-in-georgia.

Page 88. **She straddled the turtle's massive shell** Will Harlan, *Untamed: The Wildest Woman in America and the Fight for Cumberland Island* (New York: Grove Press, 2014), ix–x.

Page 90. **essential for the progress of the race** Andrew Carnegie, "The Gospel of Wealth," *North American Review*, 1889.

Page 95. **Those trees are a dramatic expression of climate change** Noe from phone interview and in Roger Drouin, "How Rising Seas Are Killing Southern U.S. Woodlands," *Yale Environment 360*, November 1, 2016.

Page 95. **The average retreat would be fifteen meters** Calhoun Daniel and Jeffrey W. Riley, "Spatial and Temporal Assessment of Back-Barrier Erosion on Cumberland Island National Seashore, Georgia, 2011–2013," USGS, 2016, https://pubs.usgs.gov/sir/2016/5071/sir20165071.pdf.

Page 96. **vignette of primitive America** A. Starker Leopold et al., "The Goal of Park Management in the United States," in *Wildlife Management in the National Parks*, National Park Service, 1963.

Page 96. **continuous change that is not yet fully understood** National Park System Advisory Board, "Revisiting Leopold: Resource Stewardship in the National Parks," 2011, www.nps.gov/calltoaction/PDF/LeopoldReport_2012.pdf.

Page 97. **70 percent of Americans say global warming is happening** Yale Program on Climate Change Communication and George Mason Center for Climate Change Communication, "Climate Change in the American Mind," May 2017.

Chapter 6. The Octopus in the Basement: Surreal Matters in the Sunshine State, Florida

Page 101. **Scientists continue to disagree about the degree and extent of global warming** Scott Pruitt and Luther Strange, "The Climate-Change Gang," *National Review*, May 17, 2016.

Page 102. **blasted off from Cape Canaveral atop an Atlas rocket** Barack Obama, Facebook post at "POTUS 44," December 8, 2016.

Page 103. **plumping them up like a fading star injecting collagen** Pittman Craig, "As Rising Sea Level Chomps at Cape Canaveral, NASA Uses Nature-friendly Solution," *Tampa Bay Times*, December 29, 2014.

Page 104. **NASA should be focused primarily on deep space** Robert S. Walker and Peter Navarro, "Trump's Space Policy Reaches for Mars and the Stars," October 19, 2016, http://spacenews.com/trumps-space-policy-reaches-for-mars-and-the-stars/.

Page 104. **The greenhouse effect has been detected** Michael Weisskopf, "Scientist Says Greenhouse Effect Is Setting In," *Washington Post*, June 24, 1988.

Page 104. **tearing out the smoke detector** Bill McKibben, "It's Time to Stand Up for the Climate—and for Civilization," January 18, 2017, www.wired.com/2017/01/stand-up-for-the-climate-and-civilization/.

Page 105. **I'm a business guy. I'm a solutions person** Mitch Perry, "Rick Scott (Sort of) Discusses His Talk with Climate Scientists," *Creative Loafing Tampa Bay*, August 20, 2014.

Page 105. **made us an economic powerhouse** Barack Obama, "Farewell Speech," January 11, 2017, *New York Times*, www.nytimes.com/2017/01/10/us/politics/obama-farewell-address-speech.html.

Page 107. **Miami is the most exposed city in the world in terms of property damage** "Climate Change Could Triple Population at Risk from Coastal Flooding by 2070," April 12, 2007, www.oecd.org/general/climatechangecouldtriplepopulationatriskfromcoastalfloodingby2070findsoecd.htm.

Page 108. **up to *four times* more carbon per hectare** USDA Forest Service, "Mangroves among the Most Carbon-Rich Forests in the Tropics; Coastal Trees Key to Lowering Greenhouse Gases," *Science Daily*, April 5, 2011.

Page 109. **land where he kept caged alligators** Craig Pittman, "The Sunshine State Becomes the Surreal State during a Weird 2016," *Tampa Bay Times*, December 25, 2016.

Page 110. **butterfly attracting vines placed along the park's perimeter** Monarch Hill Renewable Energy Park, "Letter to Coconut Creek Residents," December 30, 2014, http://monarchhill.wm.com/index.jsp.

Page 111. **fabulous muck** In Michael Grunwald, *Swamp: The Everglades, Florida, and the Politics of Paradise* (New York: Simon and Schuster, 2006), 137.

Page 112. **A sea level rise work group of the compact recommends** Sea Level Rise Work Group, "Unified Sea Level Rise Projection," South Florida Regional Compact, October 2015, www.southeastfloridaclimatecompact.org/wp-content/uploads/2015/10/2015-Compact-Unified-Sea-Level-Rise-Projection.pdf.

Page 116. **he wrote in a policy paper** Harold R. Wanless, "The Coming Reality of Sea Level Rise: Too Fast Too Soon," November 2014, Department of Geological Sciences, University of Miami, www.bio.miami.edu/arboretum/wanless.pdf.

Page 116. **West Antarctic is kind of an OMG thing** Don Jergler, "RIMS 2016: Sea Level Rise Will Be Worse and Come Sooner," *Insurance Journal*, April 12, 2016.

Page 117. **rolling easements** Stan Cox and Paul Cox, "A Rising Tide." *New Republic*, November 8, 2015.

Page 119. **an impermeable membrane** Elizabeth Kolbert, "The Siege of Miami," *New Yorker*, December 21, 2015.

Page 123. **reef system is now covered with living coral** Chris Mooney, "The Race to Save Florida's Devastated Coral Reef from Global Warming," *Washington Post*, June 25, 2017.

Page 127. **five million birds were being killed** "Everglades National Park," PBS, www.pbs.org/nationalparks/parks/everglades/.

Page 127. **keep the planet** Michael Grunwald, "A Rescue Plan, Bold and Uncertain," *Washington Post*, June 23, 2002.

Chapter 7. Springing Back: Resiliency on the Gulf Coast, Louisiana and Texas.

Page 130. **spiritual resilience** Tom Piazza, "Why New Orleans Matters," *Washington Post*, February 24, 2006.

Page 130. **The storm killed more than 1,800 people** Kevin Loria, "10 Years after Hurricane Katrina, Here's Why New Orleans Still Faces Disaster," *Business Insider*, August 29, 2015.

Page 131. **ineffable quality that allows some people to be knocked down** "All About Resilience," *Psychology Today*, www.psychologytoday.com/basics/resilience.

Page 134. **1,800 square miles** *2017 Coastal Master Plan*, June 2, 2017, http://coastal.la.gov/our-plan/2017-coastal-master-plan/.

Page 134. **We harnessed it, straightened it, regularized it, shackled it** John McPhee, "Atchafalaya," in *The Control of Nature*, (New York: Farrar, Straus and Giroux, 1990), 26.

Page 135. **declared a state of coastal crisis and emergency** John Bel Edwards, "State of Emergency—Coastal Louisiana," April 20, 2017, http://gov.louisiana.gov/assets/EmergencyProclamations/43-JBE-2017-Coastal-Louisiana.pdf.

Page 137. **tragedy of the horizon**. Mark Carney, "Breaking the Tragedy of the Horizon—Climate Change and Financial Stability," September 29, 2015, www.bankofengland.co.uk/publications/Pages/speeches/2015/844.aspx.

Page 140. **the town being built on 'made' ground** Mark Twain, *Life on the Mississippi* (New York: Harper, 1883), 300.

Page 141. **twenty-six thousand miles of pipeline throughout the southern coastal wetlands** Richard Misrach and Kate Orff, *Petrochemical America* (New York: Aperture, 2012).

Page 143. **the boot appears as if it came out on the wrong side** Brett Anderson, "Louisiana Loses Its Boot," *Matter*, September 8, 2014.

Page 143. **zone that will continue to change** Brett Anderson, "The Great Boot Debate: Is It Time for Louisiana to Change Its Map?," *Times-Picayune*, May 9, 2017, www.nola.com/futureofneworleans/2015/08/will_the_time_ever_be_right_fo.html.

Page 143. **My intention was to argue that maps** Laura Bliss, "No, Louisiana Isn't 'Losing Its Boot,'" August 24, 2016, www.citylab.com/design/2016/08/louisiana-losing-boot-map/497186/.

Page 144. **We're frickin' Indians** Anderson, "Louisiana Loses Its Boot."

Page 145. **effects of the oil and gas industry's canal dredging** Board of Commissioners v. Tennessee Gas et al., July 23, 2013, http://biotech.law.lsu.edu/blog/petition-for-damages-and-injunctive-relief.pdf.

Page 146. **We're on this earth for a limited amount of time** Arlie Russell Hochschild, *Strangers in Their Own Land: Anger and Mourning on the American Right* (New York: New Press, 2016), 54.

Page 146. **The spill makes us sad** Michael Grunwald, "Katrina: A Man-Made Disaster," *Time*, November 4, 2010.

Page 146. **The Louisiana story is an extreme example** Hochschild, *Strangers in Their Own Land*, 253.

Page 149. **Antarctica has the potential to contribute** Robert M. DeConto and David Pollard, "Contribution of Antarctica to Past and Future Sea-level Rise," *Nature*, March 31, 2016.

Page 150. **That would defeat the purpose** Ari Shapiro and Matt Ozug, "Wind Energy Takes Flight in the Heart of Texas Oil Country," March 8, 2017, https://www.npr.org/2017/03/08/518988840/wind-energy-takes-flight-in-the-heart-of-texas-oil-country.

Page 151. **trade organizations representing them may do the opposite** Sheldon Whitehouse, *Captured: The Corporate Infiltration of American Democracy* (New York: New Press, 2017), 155.

Page 153. **simple, it's conservative, it's free market** Chris Mooney and Juliet Eilperin, "Senior Republican Statesmen Propose Replacing Obama's Climate Policies with a Carbon Tax," *Washington Post*, February 8, 2017.

Page 154. **cleaving pipelines from their moorings** Roy Scranton, "When the Next Hurricane Hits Texas," *New York Times*, October 7, 2016.

Page 157. **The mournful dirges of the breakers** Austin in *Through a Night of Horrors: Voices from the 1900 Galveston Storm*, ed. Casey Edward Greene and Shelly Henley Kelly (College Station: Texas A&M University Press, 2002), 70.

Page 158. **We think it's sensible policy for the national government to be a financial partner** "Ike Dike Concept Gets a Big Backer," *Houston Chronicle*, March 22, 2017.

Page 160. **killed every one of us** David La Vere, *The Texas Indians* (College Station: Texas A&M, 2003), 62.

Page 161. **no reliable eyewitness verifies cannibalism nor does archaeology support it** La Vere, 62.

Page 162. **Range of these wild hogs is expanding** Nathan P. Snow et al., "Interpreting and Predicting the Spread of Invasive Wild Pigs," *Journal of Applied Ecology*, December 2016.

Page 163. **like a pianist playing with one hand** John McPhee, "Atchafalaya," in *The Control of Nature* (New York: Farrar, Straus and Giroux, 1990), 26.

Page 164. **warmer than average temperatures over the winter** Eric Berger, "For the First Time, the Gulf of Mexico Didn't Fall below 73° This Winter," March 1, 2017, *https://arstechnica.com/science/2017/03/a-sizzling-gulf-of-mexico-could-bring-more-spring-storms/.*

Page 164. **30 percent or so of the total rainfall** Robinson Meyer, "Did Climate Change Intensify Hurricane Harvey?," *Atlantic*, August 27, 2017.

Page 164. **according to the meteorologist Eric Holthaus** Eric Holthaus, "A Texas-size Flood Threatens the Gulf Coast, and We're So Not Ready," August 23, 2017, https://grist.org/article/a-texas-size-flood-threatens-the-gulf-coast-and-were-so-not-ready/.

Page 164. **lost almost 50 percent of its soil-rich wetlands**. "Why Is Houston So Prone to Major Flooding?," August 28, 2017, www.cbsnews.com/news/harvey-why-is-houston-so-prone-to-major-flooding/.

Page 164. **seven thousand residential buildings have been built** Nina Satija et. al., "Boomtown, Flood Town," *ProPublica/Texas Tribune*, December 7, 2016.

Page 164. **use Harvey to jump-start its transition** Vernon Loeb, "Harvey Should Be the Turning Point in Fighting Climate Change," *Washington Post*, August 19, 2017.

Chapter 8. Take in the Waters: On the Birthplace of Rivers, West Virginia

Page 166. **We're just going to thank God** "Severe Wet and Then Dry Weather Dampen Harvests," *Market Bulletin*, October 2015, www.agriculture.wv.gov/marketbulletin/Documents/10-15MB-Web.pdf.

Page 167. **ask questions to clarify evidence of the factors** Ryan Quinn, "Climate Change Learning Standards for W. Va. Students Altered," *Charleston Gazette-Mail*, December 28, 2014.

Page 171. **hillbillies don't know what's good for them** Jenna Portnoy, "After Coal, Appalachia to Wind Farm Proposal: 'It is Insulting, Really,'" *Washington Post*, August 20, 2015.

Page 173. **long term leasing of land in a rehabilitated Appalachian Commons** Ron Eller, "Deep Change the Only Path to Real, Lasting Reform in Appalachia," *Lexington Herald Leader*, February 15, 2015.

Page 179. **two-story stone gabled structure** "National Register of Historic Places Inventory Nomination Form," December 31, 1984, https://npgallery.nps.gov/pdfhost/docs/NRHP/Text/83003244.pdf.

Page 180. **a community over time and provide a quality of life.** Ron Eller in *West Virginia: A Film History*, www.wvencyclopedia.org/media/29182?article_id=1293.

Page 181. **It then went to the Smithsonian for display** Dave Tabler, "I Wish They'd a Threw It in the New River Sometimes," May 5, 2015, www.appalachianhistory.net/2015/05/i-wish-theyd-threw-it-in-new-river.html.

Page 187. **assisting wild ginseng in migration** Glynnis Board, "Wild Ginseng, Wood Thrushes, and Climate Change: A Survival Story?," October 21, 2014, http://wvpublic.org/post/wild-ginseng-wood-thrushes-and-climate-change-survival-story#stream/0.

Page 188. **technological termite-life** Wallace Stegner, *The Sound of Mountain Water* (Lincoln: University of Nebraska Press, 1985), 141.

Page 189. **I gave my heart to the mountains** Stegner, "Overture," *Sound of Mountain Water*, 42.

Page 190. **anxiety, depression, insomnia** Kai Erikson, *Everything in Its Path: Destruction of Community in the Buffalo Creek Flood* (New York: Simon and Schuster, 1976), 136.

Page 191. **days of hurt become years.** Robert Bullard and Beverly Wright, eds., *Race, Place, And Environmental Justice after Hurricane Katrina: Struggles to Reclaim, Rebuild, and Revitalize New Orleans and the Gulf Coast* (Boulder, CO: Westview, 2013), 4.

Page 191. **4 percent more moisture in the atmosphere since 1970** Kenneth Trenberth, "The Impact of Climate Change and Variability on Heavy Precipitation, Floods, and Droughts," in *Encyclopedia of Hydrological Sciences*, 2008, www.cgd.ucar.edu/staff/trenbert/books/EHShsa211.pdf.

Page 191. **71 percent increase in extreme rainfall events since 1958** National Climate Assessment, 2014, http://nca2014.globalchange.gov/report/regions/northeast.

Page 192. **the course of love is like that mountain stream** *Porte Crayon, Virginia Illustrated: Containing a Visit to the Virginian Canaan, and the Adventures of Porte Crayon and His Cousins* (New York: Harper, 1857), 257

Page 192. **We can also know that we are still a rich nation** Clinton Anderson, "This We Hold Dear," *American Forests*, July 1963, 24–25.

Chapter 9. More Ghosts on the Coasts, and the Last Place to Go

Page 194. **When the sun comes out again after the storm** Tom Dent, *Southern Journey: A Return to the Civil Rights Movement* (New York: Morrow, 1997), 148.

Page 196. **Some kind of major flood happened** Justin Gillis, "Looming Floods, Threatened Cities," *New York Times*, May 18, 2017.

Interviews and Correspondence

Alexander, Clark. Director, Skidaway Institute of Oceanography, Georgia.

Anderson, John B. Maurice Ewing Professor of Oceanography, Rice University.

Barie, Danny. Attorney at law, West Virginia.

Beever, Jim. Planner, Southwest Florida Regional Planning Council.

Bergh, Chris. Director of Coastal and Marine Resilience. Nature Conservancy, Florida.

Blackburn, Jim. Codirector of the Severe Storm Prediction, Education and Evacuation from Disaster (SSPEED) Center, Rice University.

Boyett, Ricky. Public information spokesperson, U.S. Amy Corps of Engineers, New Orleans.

Bryant, Mike. Refuge manager, Alligator National Wildlife Refuge, North Carolina.

Buelterman, Jason. Mayor, Tybee Island, Georgia.

Cabiness, Laura. Director, Public Works, City of Charleston, South Carolina.

Carney, Jeff. Director, Coastal Sustainability Studio. Louisiana State University.

Cason, Jim. Mayor, City of Coral Gables, Florida.

Jeff Chanton, Professor of Oceanography, Florida State University.

Connor, Susan. Chief, Planning and Policy Branch, U.S. Army Corps of Engineers, Norfolk, Virginia.

Costello, Mike. Director, West Virginia Wilderness Coalition.

Covi, Michelle. Assistant professor of practice, Department of Ocean Earth and Atmospheric Sciences, Old Dominion University, Virginia.

Crawford, John "Crawfish." Naturalist, University of Georgia Marine Extension Service.

Cronin, Thomas. Senior geologist, U.S. Geological Survey, Virginia.

Darby, Reginald. Legislative director, Representative Scott Taylor, Virginia.

Davis, Hamilton. Climate director, Coastal Conservation League, South Carolina.

Davis, Mark. Director, Tulane Institute on Water Resources Law and Policy, Louisiana.

DeBerry, Frank. President, Snowshoe Mountain Resort, West Virginia.

Evans, Jason. Assistant professor of environmental science and studies, Stetson University, Florida.

John Lopez, Director of Coastal Sustainability, Lake Ponchartrain Basin Foundation, Louisiana.

Everett, Christy. Hampton Roads director, Chesapeake Bay Foundation, Virginia.

Fly, Liz. Coastal climate extension specialist, South Carolina Sea Grant.

Gilmer, Ben. President, Refresh Appalachia, West Virginia.

Goodwine, Marqueta, "Queen Quet." Chieftess of the Gullah/Geechee Nation, South Carolina.

Hendricks, Stacia. Naturalist manager, Little St. Simons Island, Georgia.

Henifin, Ted. General manager, Hampton Roads Sanitation District, Virginia.

Hitt, Nathaniel. Research fish biologist, U.S. Geological Survey, West Virginia.

Inglis, Bob. Executive director, Energy and Enterprise Initiative, George Mason University, Virginia.

Ingram, Gary. Superintendent, Cumberland Island National Seashore, Georgia.

Jockman, Sebastian. Civil engineer, Delft University, Netherlands.

Jurado, Jennifer. Chief climate resilience officer, Broward County, Florida.

Kearns, Matt. Coordinator, Public Lands Campaign, West Virginia Rivers.

Maibach, Edward. Director, George Mason's Center for Climate Change Communication, Virginia.

McKinney, Linda. Director, Five Loaves and Two Fishes Food Bank, West Virginia.

McKinney, Joel. Owner, Roadside Farms, West Virginia.

Merrell, Bill. George P. Mitchell Chair in Marine Sciences, Texas A&M Galveston.

Moore, Chris. Senior scientist, Chesapeake Bay Foundation, Virginia.

Noe, Gregory. Research ecologist, U.S. Geological Survey, Georgia.

Noss, Reed. Retired conservation biologist, University of Central Florida.

Obeysekera, Jayantha "Obey." Hydrologist, South Florida Management District.

Oishi, Chris. Research ecologist, Coweeta Hydrologic Laboratory, U.S. Forest Service, North Carolina.

Parmesan, Camille. National Marine Aquarium Chair in the Public Understanding of Oceans and Human Health, Marine Institute, University of Plymouth, United Kingdom; and professor in geology, University of Texas at Austin.

Pittman, Craig. Journalist, *Tampa Bay Times*, Florida

Pruitt-Moore, Carol. Resident, Tangier Island, Virginia.

Quattlebaum, Thomas. Sea level rise fellow, Chesapeake Bay Foundation, Virginia.

Riley, Joe. Former mayor, Charleston, South Carolina.

Rodd, Tom, Project director, Allegheny Highlands Climate Change Impacts Initiative, West Virginia.

Ruckdeschel, Carol. Biologist, naturalist, Cumberland Island, Georgia.

Schleicher, Diane. City manager, Tybee Island, Georgia.

Schulte, David. Marine biologist, U.S. Army Corps of Engineers, Norfolk, Virginia.

Stacey, Charles. Supervisor, Tazewell County, Virginia.

Stewart, Dennis. Wildlife biologist, Alligator National Wildlife Refuge, North Carolina.

Stiles, Skip. Executive director, Wetlands Watch, Virginia.

Stoker, Louise. Mayor of Bramwell, West Virginia.

Traxler, Steve. Senior biologist, U.S. Fish and Wildlife Service, Florida.

Tyler, Renee. Town manager, Tangier Island, Virginia.

Wanless, Harold R. Chair, Department of Geological Sciences, University of Miami, Florida.

Wolff, Paul. Former council member, Tybee Island, Georgia.

Williams, Carolee. Project manager, Department of Planning, Preservation, and Sustainability, Charleston, South Carolina.

Index

Abbey, Edward, 44
adaptation: 6; backflow devices, 112; brook trout and, 189; controlled burns, 123; cultural, 11; as evolutionary process, 7; economic, 10, 173; evangelical Christians and, 70; grief and, 195; with oysters, 37, 48; planning, 106; pumps for sea water, 118; sea turtles, 92; trial and error, 124
aerial photographs, 94, 107, 124, 125, 149
Alexander, Clark, 81
Allegheny Highlands Climate Change Impacts Initiative, 167, 177
Alligator National Wildlife Refuge, 47–50
alligator, 50, 53–55
American Psychological Association, 191
Anderson, Brett, 143–44
Anderson, Clinton, 192
Anderson, John, 148–51, 161
Angel Oak, 194
Antal, Rev. Jim, 66
Antarctica ice, 148
Anthropocene, 165
armadillos, 60, 90
Audubon Society, 127

Baker, James, 152
Bartram, William, 54, 206n
beach renourishment, 22, 72, 103, 160
Beartown State Park, 184
Beaufort, South Carolina, 66, 72
Beever, Jim, 106–7
Beloved (Morrison), 73, 207n
Bergh, Chris, 118, 121–22, 127
Between God and Green (Wilkinson), 69
Big Chill, The, 72
Big Cypress National Preserve, 120
biocoenisis, 40
Birthplace of Rivers National Monument, 181, 183, 185, 188, 190
Blackburn, Jim, 151–55
Blue Ridge Mountains, 13, 14, 73, 78
Board of Architectural Review, Charleston, 75
Bobbitt, Phillip, 178
Boyett, Ricky, 137–38
BP oil spill, 44, 132, 146
Bramwell, West Virginia, 178–80
brook trout, 189, 199
Broward, Napoleon Bonaparte, 110
Buck, Pearl S., 184
Buelterman, Jason, 5, 84–85
Bullard, Robert, 191
Bush, George P., 155, 158
Bush, Jeb, 105

Cabiness, Laura, 62, 75, 76
Calabresi, Guido, 178
Cancer Alley, 141

Cape Canaveral National Seashore, 99, 102–3, 105
carbon dioxide, 3, 5, 45, 52, 53, 56, 67, 69
carbon sequestration, 152–53
Carnegie, Andrew, 90
Carnegie, Lucy, 89
Carney, Jeff, 143
Carol Ruckdeschel: 87–88, 91–93
Carson, Rachel, 9, 44
Cason, Jim, 5, 113–15
Chanton, Jeff, 105
Charleston, South Carolina: churches, 64–66; flood events, 12; history, 59; Hurricane Mathew and, 58, 72; investment in adaptation, 62–63; sea level rise strategy, 62, 206n
Chase, Chip, 191
Chesapeake Bay Foundation (CBF): 14, 15, 18, 19, 23, 27, 37; Brock Environmental Center, 22, 34, 41
Chesapeake Bay, 1, 14, 16, 21, 22
Civilian Conservation Corps, 53
Clean Power Plan, 101, 152
climate change commitment, 3
climate change: communication and, 50; demographic shifts, 162; denial and, 45, 69, 133–34, 153, 178; effects on poor, 3, 8, 12, 115, 120; explained, 5–6; financial institutions and, 137; mental health and, 191; religion and, 68; resistance to, 49; shift in latitude, 40; scientific certainty and, 101–2; use of the term in Florida, 106
Climate Impact Lab, 8
CNN, 32
Coalfield Development Corporation, 173–72
Coastal Conservation League, 62, 63, 71
Comprehensive Everglades Restoration Plan, 121
Coral Gables, 113–15
coral reefs, 123, 167
Corps of Engineers, U.S. Army, 30–31, 72, 85, 119, 137, 145, 159, 161, 163
Costello, Mike, 183
cottonmouth, 121, 123
Covi, Michelle, 17
Crawford, John, 79
Crayon, Porte, 192
Crisp, Charlie, 105
Cronin, Thomas, 40
Cuccinelli, Ken, 30
Cumberland Island National Park: 10, 87, 96, 195; fire management and, 89; development of, 89; history, 89; rates of erosion and, 94; wilderness, 89, 95
cypress trees, 108, 121

Davis, Hamilton, 62, 64, 71
Davis, Mark, 132, 133, 147
DDT, 9
DeBerry, Frank, 188
DeConto, Robert, 149
denial, as related to climate change, 45, 69, 133–34, 153, 178
Dent, Tom, 194
de Vaca, Cabeza, 160
Dodd, Mark, 94
Don't Even Think About It (Marshall), 45
Douglas, Marjory Stoneman, 120, 127
downscaling, 8
Dylan, Bob, 142

Edwards, John Bel, 132
Eller, Ron, 173, 179
Emerson, Ralph Waldo, 177
eminent domain, 169
Environmental Protection Agency (EPA), 101
Erikson, Kai, 190
Eskridge, James "Ooker," 27, 32
evangelical Christians, 27, 33, 69–70
Evans, Jason, 85, 105–6
Everglades National Park, 10, 110–11, 118–19
ExxonMobil, 131, 151
Eyerman, Ron, 191

False Cape State Park, 41
false spring, 2
FEMA: 18, 32, 35–6, 76, 84, 164
Fink, Denman, 113

First Landing State Park, 38
Florida Center for Investigative Reporting, 106
Florida: 5, 10; ban of the term climate change, 106; in-migration to, 100; invasive species in, 100; population growth, 110
Florida Keys, 121–23, 126
Fly, Liz, 64, 72
Folly Beach, 71
Forest Stewardship Council, 35
Fraim, Paul, 8
Frontline, 36, 68
Frost, Robert, 115
future generations, 6, 10, 41, 45, 96, 137

Galveston: hurricane of 1900, 156; renourishment, 160; sea wall, 159
Generall Historie of Virginia (Smith), 39
Genius of Earth Day, The (Rome), 44
ghost forests, 94–95, 193
Ghosts and Legends of Charleston (Roffe), 194
Gilmer, Ben, 173–74
glacial rebound, 21, 41
Glacier Bay, Alaska, 10
Glacier National Park, 10
Glass Castle, The (Walls), 175–76
Glenn, John, 102
Goldwin, Tim, 79
Goodwine, Marqueta (Queen Quet), 72–73
Google Earth, 80, 142
gopher tortoise, 118
Gore, Al, 101, 111, 122, 125
Gowdy, Trey, 68
Graham, Lindsey, 66
Great Barrier Reef, 67
grief, as adaptive trait, 195
Groundhog Day, 76
Gullah Geechee, 72–74

Haley, Nikki, 60
Hampton Roads Sanitation District (HRSD), 22
Hansen, James, 104, 116
Harlan, Will, 88
Hatfield-McCoy Trail System, 174, 179
Hayhoe, Katherine, 67, 68, 151
Hemenway, Harriet, 127
Hemingway, Ernest, 124–25
Henifin, Ted, 22
Heron, Scott, 68
Hessel, Amy, 168
Hitt, Than, 189
Hochschild, Arlie, 145–46
Holdren, John, 6
Holocene, 79, 165
Honoré, General Russell, 140, 142
Horton, Benjamin, 149–50
Houston Ship Channel, 154
Houston, Texas, 148, 162; expanding economy, 152; growth, 162; historic rainfall, 153; Hurricane Harvey, 164–65; zoning, 164, 169
Hsiang, Solomon, 8
HUD (U.S. Department of Housing and Urban Development), 133, 147
Hughes, Terence J., 196
hurricanes: Harvey, 11, 36, 159, 163–65; Irma, 11, 36, 75; Katrina, 36, 129–30, 131, 132, 133, 135, 136, 138, 139, 140, 156; Mathew, 58, 72, 80, 85, 89; Sandy, 1, 27, 36, 44, 63, 132, 157, 159

ice melt, 3, 6, 10, 12, 40, 45, 64, 104, 116, 148, 150
Ike Dike, 155–58, 160, 162
In a Barren Land (Mitchell), 39
Inglis, Bob, 5, 67–70
Ingram, Gary, 89, 96
Invasion of America, The (Jennings), 46
IPCC (Intergovernmental Panel on Climate Change), 3–4, 112, 116, 122, 150
Isaac's Storm (Larsen), 156
Isle de Jean Charles, 133, 140, 143–44, 147

Jackson, Chester, 94
Jarvis, Jonathon, 96
Jennings, Francis, 39
Jurado, Jennifer, 112, 113, 118
Justice, Jim, 182

Karankawa Indians, 160, 161
Kearns, Matt, 183–84, 188, 190

Key deer, 122, 114
Key West, Florida, 124–26, 128
Kitty Hawk, 46, 105
king tide, 9, 61, 75, 118, 123
Kiptopeke State Park, 23
Klein, Naomi, 9
Koch brothers, 84
Kolbert, Elizabeth, 119
Kroegel, Paul, 127

Lafayette River, 15, 16, 20
Lake Borgne Surge Barrier, 138
Lake Ponchartrain Basin Foundation, 134, 136, 146
Landrieu, Mitch 132
La Vere, David, 161
Lawson, John, 54
leatherback sea turtles, 83
Leopold, A. Starker, 96
Leopold, Aldo, 96
Linger, Wade, 167
Little Ice Age, 39
Little St. Simons Island, Georgia, 86, 92
Living Building Challenge, 34
living shoreline, 18, 19, 20, 37, 91, 193
Locklear III, Samuel J., 45
loggerhead sea turtles, 92, 94, 98
Lopez, John, 134, 146
Louisiana Coastal Protection and Restoration Authority (CPRA), 133
Louisiana, 134–135
Louisiana's Strategic Adaptations for Future Environments (LA SAFE), 133
Lynnhaven River NOW, 37
Lynnhaven River, 34, 35

Maibach, Ed, 198
mangroves, 107–8, 121
Mann, Michael E., 30
Mar-a-Lago, 117
Marine Education Center, University of Georgia, 79
Marshall, George, 45
Maslow, Abraham, 115
McCarthy, Cormac, 12, 14
McGraw, Jim, 187
McKibben, Bill, 104
McKinney, Joel, 172–73
McKinney, Linda, 172
McPhee, John, 88, 163
Merrell, Bill, 155–60
Miccosukee Indians, 120
Michaels, Patrick, 30, 125
Milankovitch cycles, 67–68
Mitchell, George P., 155
Mitchell, Paula Marks, 39
mitigation, 6, 70
Möbius, Karl August, 40
Moore, Chris, 14–19, 33–34, 37, 38
Morrison, Toni, 73
Mowry, Bruce, 119
Moyers, Bill, 70
Murray, Bill, 76–77

NASA, 6, 103–4
Nash, Steve, 17
National Center for Atmospheric Research, 164, 191
National Flood Insurance Program, 36
National Geographic, 4, 49
National Park Service, 90, 91, 96
National Review, 101
natural selection, 6, 90
Nature Conservancy, 40, 47–49, 118, 121, 174
Nature, 149
New Orleans, Louisiana: 7, 10, 12, 17, 119, 129–33, 136–37, 139–42, 147, 156, 157, 158, 191, 197; levee system, 134–35; population, 135; pump station, 138; surge barrier, 138–39
New Orleans Times-Picayune, 143,
New York Times, 28, 85, 125, 154
New Yorker, 88
Noe, Gregory, 94
Norfolk, Virginia: 7, 8, 9, 14, 17–18; Chesterfield Heights, 20–21; Chrysler Art Museum, 19; Hague, 19; Hermitage, 19; Mayflower Heights, 15; navy base, 16; sea level rise and, 16; subsidence and, 21

North Carolina Coastal Resources Commission, 49
Noss, Reed, 106
NPR, 36, 85

Obama, Barack, 6, 101, 102, 105, 146, 190
O'Connor, Flannery, 199
Odum, Eugene, 82
Oishi, Chris, 50–53
optimism bias, 11, 87, 97
Orestes, Naomi, 30
Outer Banks, North Carolina, 46–47
owl, great horned, 147
oysters: 24, 33–34, 40, 102, 130, 159, 161, adaptations, 37; reefs to control erosion,. 48–49; sea level rise and, 37; water quality and, 38
oyster bags, 91, 102

Paris Climate Accord, 5, 6, 33, 66, 70, 197
Parmesan, Camille, 2
Paulson, Hank, 36, 86
Pelican Island National Wildlife Refuge, 127
Perry, Rick, 150
Petrochemical America (Misrach and Orff), 141
Pilkey, Orrin, 49
piping plover, 49
Pittman, Craig, 103, 126
plastic bag ban, 83–84
Pleistocene, 79
Plum Orchard (Cumberland Island), 89
Pope Francis, 65
Poquoson, 18
Post and Courier, 63, 71
Pruitt, Scott, 101
Pruitt-Moore, Carol, 1, 25–27
Psychology Today, 131
pygmy rattlesnakes, 108–9

Quattlebaum, Thomas, 15–18, 35–38
Queen Quet (Marqueta Goodwine), 72–73

Rahall, Nick Joe, 180
renewable energy, 102, 113, 117, 171
renourishment, 22, 72, 103, 160
reptilian brain, 44, 46, 50, 55
Res/Con, 131, 137
resilience: 9, 17, 48. 56, 98; defined, 131–32
Revkin, Andrew, 125
Riggs, Stanley, 49
Riley, Joe, 62–63
risk communication, 106
Rising Tides Summit, 113, 114
Risky Business Project, 36
Road, The (McCarthy), 12, 14
Rockefeller Foundation, 17, 131, 132, 135
Rockefeller, John D., 131
Rodd, Tom, 167, 177
Roffe, Denise, 194
Rolling Stone, 117
Rome, Adam, 44
Rubio, Marco, 104, 108

salamanders, 109
Sanders, Bernie, 173
Sanford, Mark, 66
Savannah River, 83, 94
Schulte, David, 23–25
science, benefits of, 105
Scott, Rick, 105–6, 108, 113
Scranton, Roy, 154
Sea Grant (NOAA), 20, 64, 94, 106
sea level rise: causes of, 3, 40; on Cumberland Island, 94; Norfolk, 8, 16, 21; FEMA maps and, 36; Florida Keys, 126; levels during previous epochs, 79, 116, 149; maps and, 143; NASA Cape Canaveral, 112; North Carolina, 49; politics and, 8; projections, 112, 116; rate of, 6, 21, 150; relative, 21; relocation and, 73, 116, 124, 133, 144, 148; South Carolina, 66; Tangier Island, 24; terms used instead of, 5, 17; trees and, 48, 94–95, 193; Trump properties, 117
sea turtles, 88, 92–93, 98
season creep, 2
Sewell's Point, 16, 32
Sierra Club, 84
Silent Spring (Carson), 9

Skidaway Institute of Oceanography, 81
Skidaway Island, 79
Smith, John, 24, 26, 38, 39, 41
Smith's Island, 32
Smithsonian, 93
solar energy, 34, 69, 101, 112, 171, 174, 175
Southern Journey, A (Dent), 194
Southeast Florida Regional Climate Change Compact, 112
spring peepers, 43, 44, 57
spring tides, 9
Stegner, Wallace, 90, 188, 190, 192, 213n
Stiles, Skip, 19–20, 21
Stocknes, Per Espen, 7
Stoddard, Phil, 117
Stoker, Louise Dawson, 178–81
Strangers in Their Own Land (Hochschild), 145, 211n
strategic retreat, 12, 21, 73, 103, 117
subsidence: 11; glacial rebound, 21, 41; Louisiana, 134; Norfolk, 21; oil and gas extraction, 134; water drawdown and, 22
Sullivan's Island, 58, 70–71
sustainability, 131
Sweetwater, Florida, 119

Tampa Bay Times, 103
Tangier Island: 1, 5, 22–23, 27, 29; dialect, 24; history, 25–26, 99; and land loss, 24; media coverage, 28, 32; Uppards Island, 25–26
Taylor, Scott, 30
Tea Party, 62, 84
Texas Bureau of Environmental Geology, 149
Texas Commission on Environmental Quality, 150
This Changes Everything (Klein), 9
Thoreau, Henry David, 43, 44, 186, 187, 198
tidal gauges: Fort Pulaski, 84, 85; Sewell's Point, 16
tidal prism, 82
Tillerson, Rex, 101, 114
Tom Piazza, 130
Tragic Choices (Calabresi), 177
trauma, 11, 131, 190–91
Travels (Bartram), 54
Traxler, Steve, 100, 127
tree ring cores, 39, 56, 168
Trenberth, Kenneth, 164, 191
Trump, Donald, 5, 27, 32–33, 101–2, 117, 135, 160
Trust for Public Land, 34
turtle excluder devices, 93
Twain, Mark, 140
Tybee Island, 82–86
Tyler, Renee, 27–29
Twitter, 163

Udall, Stewart, 96
U.S. Geological Survey, 40, 95, 189

Virginia Beach, 33–35
Virginia Climate Fever (Nash), 17
Virginia Institute of Marine Science (VIMS), 17
Virginia Marine Resources Commission, 24
Virginia northern flying squirrel, 186

Walls, Jeannette, 175, 177
Wanless, Hal, 115–18, 126
Washington Post, 32, 50, 66, 153, 164, 171
Weather Channel, 76
West Virginia Wilderness Coalition, 183
Whitegrass Ski Area, 189, 191
Whitehouse, Sheldon 151
Wilkinson, Katharine, 69, 70
Williams, Carolee, 61, 62, 75
Williams, Ron, 17, 21
wind energy, 69, 150, 171
winter wren, 184
Wolff, Paul, 83–86
Wright, Beverly, 191

Yale University Program on Climate Change Communication, 45, 47, 97

Zika virus, 105